DELICIOUS CALCIUM-RICH DAIRY-FREE VEGETARIAN RECIPES

CalciYum!

DAVID & RACHELLE BRONFMAN

Acclaim for *CalciYum!*

"From a physician's perspective, this book fills a real need. This book guides you to the many foods that provide abundant calcium and are loaded with many other essential vitamins and minerals. It shows how to boost calcium intake and, at the same time, reduce your body's calcium losses quickly and easily. I recommend this book wholeheartedly."

NEAL D. BARNARD, M.D.
President, Physicians' Committee For Responsible Medicine, Washington, D.C.

"*CalciYum!* proves that food can be healthy and taste absolutely fabulous."

JOHN ROBBINS
Author of *Diet for A New America*

"For every person attempting to become vegetarian one of the first questions is 'where do I get my calcium?' When this question is posed to me, I answer 'read *CalciYum!*' It is impeccably researched, very readable and, most important, the recipes are delicious and good for you. There is no better source of information in print today."

HOWARD F. LYMAN
Past President, International Vegetarian Union
and celebrated author of *Mad Cowboy*

"The Bronfmans have delivered an invaluable resource for the calcium-challenged: a cookbook packed with nutrient-dense recipes that maximize taste as well as nutrition. This book deserves 'top shelf' in the kitchen of everyone who enjoys eating well."

SUZANNE HAVALA, M.S, R.D., FADA
Author and Nutrition Advisor, The Vegetarian Resource Group

"Vegetarians, those who are lactose intolerant or anyone interested in a healthier lifestyle will want this book on their kitchen shelves."

Vickie Elmore, *Healthy & Natural Journal*

"To describe *CalciYum!* as merely 'outstanding' would be understatement. The word 'seminal' is more apt. The Japanese have a still better word for such excellence: *shibui* – a thing of beauty and practicality in perfect balance."

Jim Oswald, *Plant Based Nutrition Institute*

"Weight smart recipes from *CalciYum!*"

Modern Woman magazine

"How do you get your calcium if you don't consume dairy products? The best answer yet is called *CalciYum!*"

Forever Young magazine

"Not many doctors or even dieticians have had the answers people need to go dairy-free, and never has such information been compiled in a concise, easy-to-use volume."

Lifelines, Toronto Vegetarian Association

DELICIOUS CALCIUM-RICH DAIRY-FREE VEGETARIAN RECIPES

CalciYum!

DAVID & RACHELLE BRONFMAN

Bromedia Inc.

CalciYum!
Calcium-Rich, Dairy-Free Vegetarian Recipes

For complete cataloguing data, see page 6.

Design & page composition: Matthews Communications Design
Photography: Mark T. Shapiro
Art direction/food photography: Sharon Matthews
Food stylist: Kate Bush
Prop stylist: Miriam Gee
Recipe editors: Rachelle Bronfman and Sheryl Traber
Nutritional analysis: Info Access (1988) Inc.
Indexer: Barbara Schon
Color scans & film: PointOne Graphics
Printing: The Bryant Press

Cover photo: FOUR-SEASON SAVORY STEW (PAGE 82)

Special orders for this book are available to organizations, associations and groups for fundraising, special events and promotions. For sales information, contact Bromedia Inc., 550 Eglinton Ave. W., Suite 380-77, Toronto, Ontario, Canada M5N 3A8.

Published by Bromedia Inc.
Tel: (416) 512-2965   Fax: (416) 226-0499
E-mail: info@bromedia.com

To contact the authors, write or e-mail to the address above.

Printed in Canada

First printing: July 1998
Second printing: April 1999

Contents

Canadian Cataloguing In Publication Data
Bronfman, David, 1960-
  CalciYum! : delicious calcium-rich dairy-free vegetarian recipes

Includes index
ISBN 0-9683503-0-5

1. High-calcium diet – Recipes.   2. Vegan cookery.   I. Bronfman, Rachelle. 1960-
II. Title.

RM237.56.B76 1998   641.5'632   C98-930861-8

To our daughters,

who illuminate the future

with the sparkle of their ways.

Acknowledgments

The following people have been a tremendous help. Please accept our sincere thanks:
Suzanne Havala, MS, RD, for her tireless feedback on so many nutrition-related questions. We're very grateful for her help.

Reed Mangels, Ph.D., RD, for her generous responses to our queries, and for her review of the draft manuscript.

Bonnie Kumer, RD, for her early feedback and her research contributions.

P. Mark Fromberg, M.D., for his helpful editing suggestions and medical counsel.

Charles Stahler and Debra Wasserman of The Vegetarian Resource Group.

Connie Weaver, Ph.D., RD, for her valuable feedback generally, and particularly for her early review of our Calcium Table.

Neal D. Barnard, M.D., for writing the much-appreciated foreword and for his moral support.

Chris & Sheryl Traber, for their project planning, editing assistance, recipe testing, overall guidance and moral support; and, to Paige Traber, for her patience, love and laughter.

We'd also like to thank:
Ilene Bronsteter of Bronson's China and Gifts for the fine tableware and cutlery.

Marie Claude Thibault and Carol Landry of the Canadian Produce Marketing Association.

Alison Fryer of The Cookbook Store, for her helpful overall suggestions.

David B. Haytowitz, M.Sc., of the U.S. Department of Agriculture Nutrient Data Laboratory.

Michael A. Klaper, M.D., for his inspiration to examine the advantages of plant foods over animal products.

Thanks go out to the team that made the book come alive:
Peter and Sharon Matthews; Mark Shapiro; Miriam Gee; Kate Bush; Barbara Selley; B.A., RD; Sharyn Joliat, M.Sc., RD.

We also extend our thanks to: Stephen Leckie; Kevin Pickard; Fernanda Edwards; Carol Elmalem; Bhiku Jethalal, M.D.; Ron Cridland, M.D.; Judy Chong; Rosanne Silverman; Ziona Swigart; Kelly Nurnberger; Lori Palatnik; Heather Rubinoff-Ockrant; Renata Richardson; Lawrence Sanders, M.D.; Ann Sanders, BSc.N.; Camille Dan, BSc.N.; Stacy Markin; Dar Ververka; Eleanor Yanover; Susan Melnik; Cynthia Walsh; Josie Dela Cruz; and the late Frank A. Oski, M.D.

Physician's Foreword

Many of us are rethinking the way we eat nowadays, and rightly so. To lose weight, lower our risk of cancer, heart disease, diabetes, or to feel more energetic, we're replacing chicken and pizza with pasta marinara and veggie chili, and swapping ice cream and cheese for healthier alternatives.

Many people are steering clear of dairy products in order to ease digestive troubles, joint pains, headaches, or other symptoms. While the change may well help them feel better, it brings a common worry: With milk products off the menu, where will we find calcium? In this book, David and Rachelle Bronfman answer that question with a simple and delightfully practical approach.

This book guides you to the many foods that provide abundant calcium and are also loaded with many other essential vitamins and minerals. It shows how to boost calcium intake and, at the same time, reduce your body's calcium losses quickly and easily.

From a physician's perspective, this book fills a real need. Not many doctors — or even dieticians — have had the answers people need to go dairy-free, and never has such information been compiled in a concise, easy-to-use volume. This book provides those answers with the personal touch of the authors' own experience, supported by solid research.

The recipes are original and inventive, bringing a new twist to hearty, vegetarian foods. Some include new tastes that you'll wish you had tried long ago. Others are as familiar as old family favorites.

Research has shown time and again the tremendous benefits that a better diet — rich in plant foods — can bring. Longstanding symptoms often melt away, and people start living again. This book will allow many people to discover the benefits of healthful dairy-free vegetarian foods — with a calcium boost — both for themselves and for their loved ones.

I recommend this book wholeheartedly.

Neal D. Barnard, M.D.
President
Physicians Committee for Responsible Medicine
Washington, DC

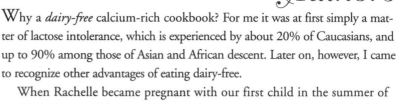

Authors'

Why a *dairy-free* calcium-rich cookbook? For me it was at first simply a matter of lactose intolerance, which is experienced by about 20% of Caucasians, and up to 90% among those of Asian and African descent. Later on, however, I came to recognize other advantages of eating dairy-free.

When Rachelle became pregnant with our first child in the summer of 1994, the issue of dietary calcium took on a new significance. We had done enough nutritional research to feel comfortable with the amount of calcium we were getting without dairy foods. But with Rachelle expecting, we began to wonder if she was getting enough calcium in her diet. Given that calcium-fortified soy milks were not allowed into Canada at that time (although they are now), there appeared to be few alternatives to the conventional wisdom that dairy products were essential as a source of calcium during pregnancy. This dilemma became especially acute after we had begun a series of pre-natal classes, during which the subject was given considerable attention. According to several sources of information, a daily calcium intake of 1000 mg (or more) was recommended during pregnancy.

The result was that Rachelle became sufficiently concerned about the possibility of calcium deficiency that she began to add some dairy to her diet. It wasn't a happy choice, and we became determined to seek alternatives. Our first step was to review the calcium research available at the time. I also began to compose a table of plant-based foods that contain calcium. And I was able to show that non-dairy foods could deliver calcium in very respectable quantities.

While this information was useful, it was just a start. Individual food ingredients are fine, especially for the fruits and nuts that are commonly eaten raw. However, that didn't help much for the vegetables, beans and grain products that are used principally in cooking. With my experience as a freelance writer, and with Rachelle's knack for composing recipes, we believed a dairy-free calcium cookbook would benefit not only ourselves but many others. Hence the genesis of this book.

I hope you enjoy using *CalciYum!* and that it provides you with the assurance you need to move towards a calcium-rich, healthier and more exciting style of eating.

David E. Bronfman
Toronto

Introduction

Since we adopted a vegetarian lifestyle in 1990, David and I have experienced tangible health benefits and discovered an entirely new realm of fabulous foods and flavors. On our quest to prepare tasty and nutritious meals, we, like all culinary explorers, found that necessity spawned invention.

Accordingly, we began creating our own recipes. As our knowledge of — and familiarity with — vegetarian foods increased, so did our confidence in using a wider variety of ingredients. So too did we increase our understanding of what these foods offered in terms of nutrition and taste. But it was due to my first pregnancy that we began to investigate calcium-rich, dairy-free foods.

As I stalked the markets for calcium-rich ingredients, I discovered I had much to learn about high-calcium grains, beans and vegetables. In fact, so unfamiliar were many of the ingredients that I began to take my camera to the grocery store — the better to help identify collards versus kale, rapini versus broccoli, Chinese lettuce versus bok choy and the difference between various types of squash. Fellow shoppers would often point quizzically to my colorful, alien haul and ask what its end purpose could possibly be.

Friends and family began asking, "how do you get your calcium if you don't eat or drink dairy products?" I responded with kale, bok choy, quinoa, blackstrap molasses, amaranth flour — the list went on and on. Inevitably, their next question was, "what's that and how do you cook with it?"

Because of the curiosity demonstrated by family and friends, our focus turned to educating others with *CalciYum!*. We developed a convenient reference chart (see page 174) which provides calcium values for a wide range of ingredients, as well as an ingredient reference (see "The Calcium-Rich Vegetarian Pantry", page 17) which describes the foods that people so often had asked us about. The most important feature of this book, however, is the collection of recipes specially developed to deliver great taste and generous amounts of calcium.

Our cookbook is the result of several years of research, exploration, trial and re-trial. It is, we hope, an exciting guide to a new world of calcium-rich, dairy-free dishes that await your discovery.

Rachelle S. Bronfman
Toronto

Calcium: Searching for the Magic Bullet

Calcium is a mineral that plays an important part in normal body function. It is found principally in our bones, which act as a storehouse. Since we're commonly told that dairy products are the best source of calcium, we must then ask whether dairy products are the best overall choice for health. If you believe that the answer is "no," then this book will help you to find non-dairy calcium alternatives. We'll also look at the lifestyle choices best-suited to maintaining strong bones, thus avoiding osteoporosis.

While calcium intake is important, it is sometimes recommended as a key to prevent osteoporosis. But unknown to most people, a host of knowledgeable health professionals say that calcium is not the magic bullet for bone strength that it's often portrayed to be. Even two known researchers in the field — Peter Burckhardt and Robert P. Heaney — readily admit that osteoporosis is a "multifactorial disorder," meaning it's thought to be caused by a number of factors. A recent study involving 12,000 women ("The Nurses Health Study") headed by Harvard University's Walter C. Willett, M.D., has looked at many diseases over a 12-year period. The study has observed that there may be no relationship between consuming very large amounts of calcium and bone strength.

It is a gross distortion to suggest that a high-calcium diet will prevent osteoporosis. Rather, calcium is only one factor in a number of lifestyle components that can help us build and maintain good bone strength. For healthy bones, we also need to reduce our protein and salt intake and increase the amount of weight-bearing exercise we do. These things require some commitment. (See Appendix D: "Lifestyle Factors and Osteoporosis," page 177.)

It's helpful to remember that plant foods containing calcium are also loaded with other important components, such as vitamins, minerals, cancer-fighting antioxidants and dietary fiber. (Fiber is entirely absent from dairy products and all other animal-source foods.) And remember that no one recipe will have enough calcium for a whole day. Instead, what we consume throughout the day will provide us with what we need. This includes all our meals, drinks, and snacks, all adding to our calcium intake. Most importantly, the wider the variety of foods we eat from the plant kingdom, the greater the assortment of nutrients that will benefit our health.

This book will not give advice on calcium intake. What we have tried to do is provide an alternative to conventional information that says we should consume dairy products in order to meet our calcium needs. Although dairy products tend to be high in calcium, we feel that the largely untold story of dairy alternatives has to be considered.

The dairy dilemma

There are many reasons why people may not choose to rely on dairy products as a source of calcium.

Lactose intolerance

A large proportion of the world's population cannot tolerate lactose, the milk sugar in dairy products. Symptoms often include stomach discomfort such as bloating, gas, pain, and diarrhea. Symptoms are sometimes disguised or ignored. Approximately 20% of Caucasians — and the majority of those of Asian, Middle Eastern and African descent — have some degree of dairy intolerance. Nowadays, drug companies are marketing medications that trick the body into accepting dairy products more readily. This goes against any common sense: if we're not biologically suited to consuming dairy products, should we be tinkering with our natural mechanisms?

Allergies

Dairy products tends to be a high-allergen food group. This is not the same as lactose intolerance. Those allergic to dairy products (actually, it's the milk protein) exhibit symptoms such as skin problems and asthma; and for some people, dairy can promote the formation of mucus. Furthermore, milk can also contribute to a variety of other symptoms which may be fairly mild and thus often overlooked. These may include nasal congestion and ear infections.

Cataracts, infertility and cancer

Some scientific research suggests that dairy products may contribute to a higher incidence of cataracts, female infertility, as well as ovarian and breast cancer.

Salt (sodium)

Dairy products can be high in sodium, which is known to be a major factor in the onset of high blood pressure. Processed cheese products, in particular, often contain a lot of sodium.

Juvenile-onset diabetes

Some scientific research suggests that cow's milk may be partly responsible for triggering juvenile-onset diabetes in infants prone to the disease. There is an overwhelming scientific consensus that breast-fed babies do far better than bottle-fed infants.

Colic

Breast-feeding mothers drinking cow's milk, and bottle-fed infants consuming cow's milk-based formula, suffer a higher incidence of colic. Medical authorities recommend that children under one year of age stay away from cow's milk. That's because cow antibodies and dairy proteins can upset the digestive system in infants.

Migraine headaches

Migraine sufferers are cursed with a lifetime affliction that can significantly impair their social and working lives. These headaches have many dietary and environmental triggers, one of which is dairy products. Since dairy is so common in Western cultures, it is not always recognized as a possible trigger. Yet, for many, it may be one of only a few variables that can be changed. For some people, the total elimination of dairy products can have dramatic results, with a significant reduction in the frequency and the intensity of migraines.

The Calcium-Rich Vegetarian Pantry

VEGETABLES

GREENS

Beet greens A dark green leaf with red stems. Once cooked, it softens and tastes much like spinach. Beet greens are higher in absorbable calcium than spinach, so it's a good substitute. The red leaves stain whatever is cooked with them but makes an exciting visual change to the recipe. Nice for juicing with apples: yields a nutrition-packed grape-juice-like beverage (see beverage recipes).

Bok choy A type of Oriental cabbage with dark green leaves and white stalks. Popular in Chinese cooking, it has a mild flavor and can be used in a wide variety of recipes, including soups, casseroles or stir-fries. Leaves cook very quickly; for stir fries, add them toward the end of cooking. Refrigerate unwashed in a plastic bag for up to 4 days.

Chinese cabbage (Napa) A cylindrical, pale green cabbage with rippled leaves. This subtle vegetable can be used raw in salad or coleslaw, stir-fried or added to a variety of other recipes such as soups, casseroles and burgers. Similar in appearance to Chinese Lettuce, only somewhat darker in color and considerably stubbier. Choose firm, fresh leaves. Store in the fridge, wrapped in paper towel, for up to several days.

Chinese lettuce (Long Napa) Tall, pale green, with slightly rippled leaves, this can be used interchangeably with Chinese cabbage (which it resembles, although lighter in color, considerably longer, and straighter). Choose firm, fresh leaves. Store in the fridge wrapped in paper towel for up to several days.

Collard greens A type of cabbage, but looks more like a Romaine lettuce with its bunches of dark green leaves. Mild enough to be used raw in salads but the firm leaves should be shredded and lightly steamed to soften slightly. Often more appealing when cooked and blended into recipes. Choose fresh-looking green leaves that are crisp and unwilted. Wash in sink full of water and rinse under tap.

Dandelion greens A bitter, leafy green vegetable with a tough stem that can be used in soups, casseroles and salads. The long, slender, fine green leaves are semi-curled. Add these greens to strong-flavored stews, casseroles, soups and other recipes. Simmering in a little water for 7 to 8 minutes with the lid partially off helps to reduce the bitter taste. Before adding these raw greens to a salad, it's best to remove part of the tough stem. Look for fresh green color; leaves should not be yellowed. Wrap unwashed greens in damp paper towel and refrigerate in crisper for up to 5 days.

Kale Typically available in the dark green variety, kale has a large, curly leaf; bunched together, it resembles an overgrown parsley. Kale has a slightly sharp flavor with a tough leaf. Best steamed or blanched to soften the leaves. Blanching will produce a milder taste. Can be easily added to soups, casseroles, burgers and many other dishes without blanching first. Wash well in sink full of water. Discard stems and rinse under the tap. Shrinks somewhat with cooking, although not as much as many other greens. Before adding to a salad, shred, then steam until slightly soft. Easy to grow and can survive harsh winter climates.

Mustard greens A rough, light-green leafy vegetable grown principally in the Far East, where it is popular for pickling. Use it in strong-flavored soups, stews, casseroles and other cooked recipes. Discard stems and rinse under water before using. Like all leafy green veggies, it shrinks dramatically with cooking, so try not to underestimate the amount required. Look for crisp young leaves with a rich green color. Store in plastic bag in fridge for up to 1 week.

Rapini (broccoli raab) While this firm, leafy-green vegetable is from the same family as broccoli, it is considerably more bitter. It looks similar to broccoli (bunched with thick stems) but it has smaller flowering heads and is much leafier. Trim off the very bottom of the stems, though all parts of the plant are edible. The leaves tend to be the most bitter. Has often been associated with Italian cooking but is increasingly being used in a variety of recipes by chefs in North America. Not suitable for eating raw. Avoid yellowed or wilted stems and leaves.

Turnip greens The leaves of the turnip are an inexpensive green vegetable, often used in stews and soups. Cooking is almost always recommended so as to temper its slightly bitter flavor. Either add directly to a recipe, or steam or boil for several minutes. Keep pot partially uncovered when cooking to allow escape of the volatile acids (a sulphur compound), which produce the bitterness.

Handling and storage of greens
Do not wash until ready to use. Wrap greens in a damp paper towel or clean damp cloth. Place inside a plastic bag. Do not close plastic bag tightly. Simply twist the open end to close. Do not use twist-tie. Place plastic bag inside vegetable drawer or crisper in refrigerator. Greens keep fresh from a couple of days up to 1 week. Wilted and/or yellowed leaves indicates lack of freshness. Best used as soon as possible. Fresh greens lose about three-quarters of their volume when cooked.

OTHER CALCIUM VEGETABLES

Broccoli A variety of cauliflower that was developed in the Italian province of Calabria; in some areas of Europe it is still referred to as Calabrese. In North America, broccoli started to become widely available in the 1920s, following an Italian migration to California. Look for deep colored, crisp, firm plants carrying a mild, sweet smell, preferably with thinner stems. Trim off bottom of stems. Peel stems if using. Store in refrigerator in an open or perforated plastic bag. Keeps up to 5 days. Frozen broccoli is often a convenient alternative to fresh.

Brussels sprouts Originally grown in large quantities near the Belgian city of Brussels during the 16th century, Brussels sprouts are a member of the cabbage family and, in fact, resemble miniature cabbages. Look for firm, bright green product. The leaves should be tightly closed. To store, discard spoiled leaves and place in a plastic bag. Keep in refrigerator for up to 5 days or so. Like most fresh vegetables, they're best consumed as soon as possible.

Okra This green, velvety pod vegetable — used widely in Creole cooking — produces a gooey, seed-filled liquid when cooked. Can be stir-fried on its own, or used in recipes. Wash just before cooking. Dry after washing and trim ends. Look for deep green, firm pods. Note: brass, aluminum, iron or copper cookware tends to discolor okra, though this does not affect the flavor. Store in fridge for up to several days wrapped in paper towel and placed inside plastic bag.

Parsnips Similar in appearance to carrots but white/tan in color and higher in calcium. Offers a mild aroma with a sweet, almost-nutty flavor. In recipes, can often be used in place of carrots. Store in the fridge, just as you would carrots — for up to 2 weeks.

Rutabaga Sweet yellow waxy-looking root vegetable. Before using in a recipe, rutabaga should be peeled, boiled and mashed. Excellent addition to hearty cooked recipes, such as stews. Store in the fridge for up to 2 weeks.

Squash (Summer) Considered part of the cucumber family and, until quite recently, was popular mainly in Mediterranean cooking. Includes varieties such as zucchini (the most popular), golden zucchini, yellow crookneck and yellow straightneck. All varieties can be used interchangeably. Can be enjoyed both raw or cooked. Peel or scrub skin under tap and remove ends. Store in fridge in unsealed plastic bag for up to 4 days.

Squash (Winter) Was a staple of the North American Indians for thousands of years. Includes varieties such as kabocha, acorn, butternut, buttercup, spaghetti and pumpkin. Nutritionally similar. Store in a cool spot for 1 month or more.

Sweet potato This Central and South American root vegetable is a member of the morning-glory family, whereas common potatoes are from the nightshade clan. Similar to a large baking potato, but with tapered ends. Use as you would regular white potatoes. The darker, orange-skinned ones tend to be sweeter. To

bake in their skins, plain, allow about 1 1/2 hours at 425° F (220° C). Store in the open or in a dark place at room temperature for up to 1 week.

Yellow (Wax) and Green (Snap) beans Thought to have originated from the Andes mountains in South America, these long beans are eaten whole. The green and yellow varieties can be used interchangeably. Look for ones which snap easily. Best when lightly steamed or cooked. Store in a plastic bag in the fridge for up to 4 days.

GRAINS AND FLOURS

Amaranth On its own, this grain has a nutty, corn-like flavor. By far the most calcium-rich of grains, these tiny seeds can be used in recipes, as a side dish, or purchased in flour form to be used in baking, providing a nutty flavor. The whole grain requires a very fine strainer to rinse.

Carob flour Produced from the pod of a Mediterranean evergreen tree and sometimes referred to as "St. John's Bread" (or "Bokser" in Hebrew), carob resembles cocoa in flavor — but less sweet and contains no caffeine. Works nicely in baked goods and in non-dairy shakes.

Quinoa Pronounced "keen-wah," this small South American, highly nutritious, round grain was once a staple of the Incas and is still commonly used in several South American countries. It cooks quickly and has a slightly bitter flavor. Quinoa has made a comeback in recent years and can often substitute for more conventional grains. Cooked, the germ of the grain unfolds, revealing a spiral "tail."

Cooking whole-grain quinoa or amaranth

It's far easier to measure whole grains when they are dry. Measure 1 3/4 cups (450 mL) water for every 1 cup (250 mL) quinoa or amaranth grain.

Place measured quinoa or amaranth in a large bowl with plenty of fresh water. "Scrunch" the grain in your fingers by repeatedly "grabbing" and releasing it. Discard the discolored water through a fine strainer and repeat until the water comes clean.

Place measured water and quinoa or amaranth into cooking pot. On stovetop, bring to full boil, covered, then reduce heat to simmer. Continue simmering for 15 minutes. Remove from heat, uncover, and allow to cool slightly before serving.

Soy flour A fine, cream-colored flour made from milled soybeans (unlike most flours which are made from whole grains). Usually works best in recipes when combined with other flours. Choose the "defatted" variety for its additional calcium and its lower fat content.

Cooking tips for other grains

Brown rice With a strainer, rinse rice under cold running water. In a small pot with a lid, combine 1/3 cup (75 mL) rice and 1 1/3 cups (325 mL) water; bring to a boil. Reduce heat, cover and simmer for 40 to 45 minutes or until all water has absorbed. Set aside.

Millet With a fine strainer, rinse 1 cup (250 mL) dry millet under cold water. Add with 2 cups (500 mL) water to a medium-sized pot with a lid; bring to a boil. Reduce heat and simmer, covered, 20 minutes or until grain is soft and water is absorbed. Set aside.

Storage of grains and flours
Grains and flours are best kept in an airtight glass container, and stored in the fridge or other cool area — ideally with a temperature of 60° F (15° C) or less.

LEGUMES AND DRY BEANS

Adzuki (Azuki/Aduki) beans A small, reddish-brown bean with a somewhat nutty flavor. Commonly used in Far Eastern, as well as macrobiotic, cuisine.

Black (Turtle) beans A small, black bean often used in Caribbean and South American dishes, such as black bean soup and black bean chili. Higher in calcium than many other beans.

Garbanzo beans (Chickpeas) Mediterranean in origin, grown commonly in Asia and the Middle East, these beans are the principal ingredient in hummus. Often found in North African dishes such as couscous, these large, round, cream-colored beans have risen in popularity during recent years.

Kidney beans Native to the Caribbean and North America, this bean was a staple of the American native people. Nowadays best known for use in chili and stew dishes, this oval-shaped, red- or white-colored, fast-cooking bean is rich in taste and appearance.

Lima beans Native to the United States, these are also known as "butter beans." They are large, oval-shaped, fast-cooking and versatile.

Pinto beans Latin American in origin, spotted tan and brown in appearance, pinto beans are often used in bean salads or added to other recipes.

Refried (Pinto) beans A twice-cooked (or more) pinto bean mash, commonly associated with Mexican cooking. Used as a dip or added to other dishes.

Soybeans The soybean is a round, cream-colored, pea-sized legume which has been used in traditional Oriental cooking for thousands of years. These beans are often used to make other foods — such as tofu, soy sauce, soy milk, soy yogurt and *miso* paste. Soybeans can be used like most other beans but require longer cooking time. In addition to their good calcium content, soybeans offer a fair amount of iron — more than from most other plant-source foods.

White, Navy, Haricot, Great Northern beans Several types of white-colored beans come under the umbrella of white beans. Often found in traditional European and Middle Eastern dishes, but popularized in the U.S. by the H.J. Heinz Company as canned *Baked Beans*, these white, oval-shaped beans are often prepared in tomato sauce and molasses.

Storage of legumes and dry beans

With the exception of refried beans (a prepared food), legumes/dry beans are best stored in an airtight glass container in the fridge or other cool area, ideally with a temperature of 60° F (15° C) or less.

COOKING CHART

| BEAN OR GRAIN (1 CUP [250 mL]) | WATER | COOKING TIME | | PRESSURE COOKER* | YIELD |
| | | STOVETOP | | | |
		SOAKED	UNSOAKED		
Adzuki beans	3 1/3 cups (825 mL)	45 min	1 1/4 hrs	10 min	3 cups (750 mL)
Black beans	4 cups (1 L)	1 hr	1 1/2 hrs	15 min	3 cups (750 mL)
Chickpeas (Garbanzo)	4 cups (1 L)	2 1/2 hrs	3 hrs	30 min	2 1/2 cups (625 mL)
Great Northern beans	3 1/2 cups (875 mL)	1 3/4 hrs	2 hrs	12 min	2 1/4 cups (550 mL)
Kidney beans	3 cups (750 mL)	1 1/2 hrs	2 hrs	20 min	2 cups (500 mL)
Navy beans	3 cups (750 mL)	45 min	1 1/4 hrs	8 min	2 2/3 cups (650 mL)
Pinto beans	3 1/2 cups (875 mL)	2 hrs	2 1/4 hrs	15 min	2 cups (500 mL)
Soybeans (soak 24 hrs)	5 cups (1.25 L)	3 hrs	3 1/4 hrs	25 min	2 1/2 cups (625 mL)
White beans	3 1/2 cups (875 mL)	1 1/2 hrs	1 3/4 hrs	25 min	2 cups (500 mL)
Whole-grain quinoa**	1 3/4 cups (425 mL)	15 min	n/a	n/a	3 cups (750 mL)
Amaranth grain**	1 3/4 cups (425 mL)	15 min	n/a	n/a	2 cups (500 mL)

*   Times for pressure cooking are approximate and assume beans are soaked.

**   No soaking is required for grains.

Cooking legumes and dry beans

Sort. Examine beans spread out on a baking pan to ensure there are no stones.

Soak. This helps to eliminate the sugars that tend to cause flatulence. Soak overnight (6 to 8 hours; 24 hours for soybeans), ensuring there is ample water to cover the beans. If you cannot afford the time to soak beans, allow an extra 30 minutes or so of cooking time (see chart, above).

Rinse. After soaking, discard water and rinse thoroughly with fresh water.

Cook. On the stovetop, bring to a boil, reduce heat and simmer covered on low heat. In a pressure cooker, simmer fully pressurized on low heat.

Rinse again. Discard discolored water and rinse beans with fresh water.

MISCELLANEOUS

Tahini (sesame butter) A rich spread made from ground hulled sesame seeds. Often used in Middle Eastern cooking to make dips and sandwich spreads. Used liberally in sauce for falafel sandwiches. (Note: the calcium absorbability of sesame products remains under debate. See "Nutritional Notes on Calcium," page 171.)

Almond butter A rich spread made from whole ground almonds. Works well as a substitute for peanut butter. **Storage tip:** All nut and seed "butters" are best kept in the fridge, unless consumed soon after opening container.

Tofu (bean curd) A white, soft, soybean product made from curdled soy milk, tofu is thought to have first been used in China about 2000 years ago. Between 200 and 700 A.D., Buddhist missionaries brought the soybean to Japan. Since those ancient times, tofu has remained an important part of Chinese, Japanese, Thai and Korean cuisine, and has in this century become an important part of the Western vegetarian diet.

There are several varieties of tofu, differentiated largely according to their consistency: soft, regular, firm, or extra firm. All four types are available either in regular (or "cotton") form and commonly sold in plastic containers in your supermarket's fresh-vegetable section, or "silken" tofu, which is often found in health food shops and comes in Tetra Pak® aseptic boxes. Silken tofu tends to be smoother and creamier than cotton tofu. As tofu-making is more of an art than a science, no two brands will produce exactly the same consistency.

Soft tofu is best suited to fruit smoothies, salad dressings, desserts, sauces, dips and purées, while regular or firm tofu works well in tofu "cheesecake," casseroles, and scrambled tofu. Extra-firm tofu is your best choice for stir-fries, soups, salads, as well as for grilling on the barbecue.

The calcium content of tofu varies considerably. Look for tofu processed with either calcium chloride or calcium sulphate; check the ingredients and nutritional information on the package. If the calcium is identified only as a "percentage of daily value," you can estimate the calcium content quite easily. Assume that "daily value" is about 1000 mg. If the percentage of daily value is listed at, say, 15%, then you'll know that the tofu contains about 150 mg of calcium per serving (15% of 1000 = 150). **Storage tip:** Regular ("cotton") tofu must always be kept refrigerated, whereas unopened silken tofu is fine in the cupboard. Once opened, either type must be used immediately or otherwise stored in fresh water and placed in the fridge, changing the water daily. Tofu can be frozen for up to many months. Simply rinse, wrap, and place in freezer. Remember to check expiry date before purchasing tofu.

Blackstrap molasses A thick, black, strong-tasting byproduct of the sugar-refining industry. Used as a sweetener or a flavor-enhancer. Once container is opened, molasses is best stored in the refrigerator. (Note: regular molasses is **not** high in calcium.)

Figs Figs are best known in their dried form, although fresh figs can be found from time to time. In this book, we use dried figs unless otherwise noted. Figs have the highest calcium content of any known fruit. They're also available in many varieties. Most supermarket produce sections carry dried figs. Look for varieties such as *Mission* and *Calimyrna*. A few tips about figs: Soaking overnight renders them moist; drinking water or other beverage with figs helps digestion; remove stems before using in recipes; and, figs are best stored in the fridge.

Nutritional Analysis Notes

Computer-assisted nutrient analysis of the recipes in this book was performed by Info Access (1988) Inc., using the nutritional accounting system component of the CBORD Menu Management System. The nutrient database was the 1997 Canadian Nutrient File supplemented with documented data from USDA Nutrient Database SR11-1 and other reliable sources. The analysis was based on:

- imperial measures and weights;
- the smaller number of servings when there was a range;
- the smaller ingredient quantity when there was a range; and
- first ingredient listed when there was a choice.

Calcium-fortified soy milk and orange juice (300 mg calcium per 1 cup [250 mL]) were used in the analyses. Cooked legumes used in recipes were prepared without salt. Optional ingredients and ingredients in unspecified amounts were not included in the analysis.

Nutrient values have been rounded to the nearest whole number (with the exception of saturated fat, which has been rounded to one decimal place).

Spreads & Dips

Whether you're looking for a tasty snack or starter, here you'll find a delicious variety of calcium-rich pâtés, spreads and fillers. Our substitutes for tuna and egg salad are also delicious spread over fancy crackers, bread or stuffed into mini pitas.

Try the hummus, kale and black bean dips to make a delectable party tray. Or for a creative touch, you can buy small pre-made pastry shells and fill them with the maple potato spread or a pâté and serve warmed as fancy hors d'oeuvres.

You can create an attractive, tasty, calcium-rich luncheon buffet by serving a variety of these recipes with fancy breads and fresh veggies. These flavors, colors and textures will be much appreciated by your family and guests.

Tahini Hummus Dip

MAKES 1 1/2 CUPS (375 mL)

1 cup	cooked soybeans or white beans or 1 can (14 oz [398 mL]) rinsed and drained	250 mL
1 tbsp	olive oil	15 mL
1/4 cup	lemon juice	50 mL
2	cloves garlic, minced	2
1/3 cup	tahini	75 mL
1/4 cup	water	50 mL
1/2 tsp	salt	2 mL
	Paprika	
	Parsley sprigs	

1. In a food processor combine all ingredients except paprika and parsley. Purée until mixture has the consistency of a smooth paste. Transfer to a decorative bowl. Sprinkle paprika around the edges of dip and garnish with sprigs of parsley in the center.

This excellent dip goes well with crackers, spread on pita or bread, or as a dip for raw vegetables.

CALCIUM FACTS

Virtually all (99%) of our calcium supply is stored in our bones. On average, starting at age 40, we lose 0.5% of our bone mass each year. After menopause, women lose 1 to 2% each year. By age 60, a woman can have lost up to 40% of her bone mass.

				PER 2 TBSP (25 mL)					
Calcium	Calories	Protein	Fat	Saturated Fat	Carbohydrates	Fiber	Iron	Sodium
25 mg	74	4 g	6 g	0.8 g	3 g	1 g	1 mg	97 mg

Hummus Bean Blend

MAKES 3 CUPS (750 mL)

2 cups	cooked chickpeas, or 1 can (19 oz [540 mL]) rinsed and drained	500 mL
1 1/2 cups	cooked navy beans or 1 can (14 oz [398 mL]) rinsed and drained	375 mL
1/3 cup	lemon juice	75 mL
3	cloves garlic, minced	3
2 tbsp	tahini	25 mL
2 tbsp	water	25 mL
1/2 tsp	salt	2 mL
2 tbsp	chopped parsley	25 mL

1. In a food processor combine all ingredients except parsley; process until smooth.

2. Add chopped parsley; blend for a few seconds to create a speckled appearance. Transfer to a decorative bowl. Serve at room temperature or cool in refrigerator for 1 hour.

Creamy and rich, this hummus can be eaten alone, used as a sandwich spread or served as a thick vegetable dip. The finely chopped parsley provides color and visual appeal.

TIP

Garnish with fresh chopped parsley around the edges; or, for an unusual touch, decorate with some whole beans and chickpeas.

PER 2 TBSP (25 mL)

Calcium	Calories	Protein	Fat	Saturated Fat	Carbohydrates	Fiber	Iron	Sodium
18 mg	47	2 g	1 g	0.2 g	7 g	2 g	1 mg	50 mg

Spicy Bean Pâté

MAKES 1 1/2 CUPS (375 ML)

Try this thick and spicy pâté as a delicious sandwich spread or as a topping for fancy crackers. For added spice, simply add more cayenne pepper to taste. For those who prefer less heat, leave out the cayenne pepper — it will still taste great.

2 cups	cooked navy beans or *1 can (19 oz [540 mL]) rinsed and drained*	500 mL
1 1/2 tbsp	olive oil	20 mL
1	clove garlic, minced	1
1 1/2 tsp	soy sauce	7 mL
1/4 tsp	cayenne pepper	1 mL
1/4 tsp	salt	1 mL
2 tbsp	water	25 mL
1 tsp	paprika	5 mL

1. In a food processor purée beans until smooth.

2. In a nonstick skillet, heat oil over medium-high heat. Add garlic, soy sauce, cayenne pepper and salt; cook for 1 minute. Add puréed beans and water; stir until well mixed. Cook 2 minutes, stirring constantly, until warmed through. Serve hot or cold on crackers or bread with paprika sprinkled over top.

TIP

Navy beans can always be substituted for other high-calcium beans such as black, white or soybeans — your dip will be just as delicious!

| | PER 2 TBSP (25 mL) | | | | | | | |
Calcium	Calories	Protein	Fat	Saturated Fat	Carbohydrates	Fiber	Iron	Sodium
22 mg	60	3 g	2 g	0.3 g	8 g	2 g	1 mg	91 mg

Creative Kale Dip

Makes 2 2/3 cups (650 mL)

4 cups	packed finely chopped kale, ribs removed	1 L
1 cup	tofu mayonnaise	250 mL
8 oz	regular tofu or soft tofu	250 g
2 tbsp	lemon juice	25 mL
3	green onions, chopped	3
1	package onion soup mix	1
3 tbsp	finely chopped dill	45 mL
1	small, round loaf crusty bread	1

1. Steam or boil kale until wilted; let cool. With your hands, squeeze out excess liquid. Set aside.

2. In a food processor, combine mayonnaise, tofu, lemon juice, green onions and soup mix; blend until creamy. Add kale; process on and off for a few seconds until well mixed but without liquefying the kale. Transfer to a medium-sized bowl. Add dill and mix together.

3. With a sharp knife, cut a round hole in the top of the crusty bread loaf. Hollow out inside by tearing out pieces of bread and setting them aside. Fill with dip and serve with reserved bread pieces.

Spinach dip is great to serve at parties. In this recipe, we use kale instead of spinach since kale is much richer in absorbable calcium. The tofu provides a creamy texture and the end result is a delicious and familiar dip that everyone can enjoy.

TIP

If you prefer, refrigerate dip for 1 hour before adding to hollowed-out bread.

PER 2 TBSP (25 mL)								
Calcium	Calories	Protein	Fat	Saturated Fat	Carbohydrates	Fiber	Iron	Sodium
41 mg	48	2 g	3 g	0.2 g	4 g	1 g	1 mg	205 mg

Guacamole

MAKES 4 CUPS (1 L)

2	ripe avocados	2
4 oz	regular tofu or soft tofu	125 g
2 oz	soft tofu	50 g
1/4 cup	tofu mayonnaise	50 mL
1 tbsp	lemon juice	15 mL
2	tomatoes, seeded and finely chopped	2
1	clove garlic, minced	1
2 tbsp	minced onions	25 mL
1/4 tsp	chili flakes	1 mL
1/2 tsp	salt	2 mL
1/8 tsp	black pepper	0.5 mL

Our calcium-rich version of traditional guacamole is light and simple to prepare. It tastes great and looks quite attractive garnished with a few small tomato slices and sprigs of parsley.

1. With a sharp knife, cut avocados in half; remove pits. Scoop out avocado flesh into a bowl. Mash and set aside.

2. In a food processor or blender, combine regular and soft tofu, mayonnaise and lemon juice. Blend until smooth. Add to mashed avocado and mix until combined.

3. Add tomatoes, garlic, onions, chili flakes, salt and pepper to avocado mixture; stir until well mixed. Serve immediately at room temperature or chill in refrigerator for 1 hour and serve cold.

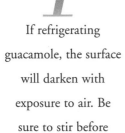

TIP

If refrigerating guacamole, the surface will darken with exposure to air. Be sure to stir before serving.

PER 1/4 CUP (50 mL)

Calcium	Calories	Protein	Fat	Saturated Fat	Carbohydrates	Fiber	Iron	Sodium
24 mg	67	2 g	5 g	0.8 g	4 g	1 g	1 mg	118 mg

Maple Potato Spread

MAKES 2 CUPS (500 mL)

1	large sweet potato, peeled and quartered	1
1	small onion, peeled and quartered	1
1 tbsp	tahini	15 mL
1 tbsp	maple syrup	15 mL
3	dried figs, finely chopped	3

1. Steam sweet potato and onion pieces until soft.

2. Transfer to a food processor with remaining ingredients. Blend until mixture is thick and smooth. Pour into a small container. Keep covered and refrigerate until ready to serve.

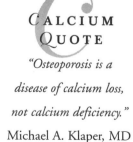

This light, sweet potato spread can be served to guests as a dip for crackers or raw vegetables. Spread it over bagels, bread or use to stuff mini pitas. You can also create a delicious and unique meal for children by stuffing it into pre-made mini pastry shells and serving them warm. This is one way we get our daughter to eat sweet potatoes and figs!

CALCIUM QUOTE

"Osteoporosis is a disease of calcium loss, not calcium deficiency."
Michael A. Klaper, MD

PER 2 TBSP (25 mL)

Calcium	Calories	Protein	Fat	Saturated Fat	Carbohydrates	Fiber	Iron	Sodium
13 mg	41	2 g	1 g	0.1 g	9 g	1 g	0.2 mg	3 mg

Liverless Liver Pâté

MAKES 1 1/2 CUPS (375 mL)

If you love the taste of chopped liver, this bean and carrot spread is unbelievably close to the real thing. Serve on bread or crackers, warm or cold.

1	large carrot, diced	1
1 cup	cooked black beans or 1 can (14 oz [398 mL]), rinsed and drained	250 mL
2 tbsp	water	25 mL
1 tbsp	canola oil	15 mL
1 cup	finely chopped onions	250 mL
1/4 tsp	salt	1 mL

1. Steam or boil carrots until soft. Set aside.

2. In a food processor, purée beans and water to form a paste. Add carrots; process until well blended.

3. In a nonstick skillet, heat oil over medium heat. Add onions and cook until soft and browned. Stir in salt and bean mixture; cook 3 minutes or until heated through. Cool, then refrigerate until ready to serve.

CALCIUM REPORT

A 1994 study published in the *American Journal of Clinical Nutrition* suggests that eliminating all animal protein from the diet cuts calcium losses by 50%.

CLOCKWISE FROM UPPER LEFT: MAPLE POTATO SPREAD (PAGE 31); ➤
CREATIVE KALE DIP (PAGE 29); GUACAMOLE (PAGE 30)

PER 2 TBSP (25 mL)

Calcium	Calories	Protein	Fat	Saturated Fat	Carbohydrates	Fiber	Iron	Sodium
27 mg	77	3 g	2 g	0.2 g	11 g	3 g	1 mg	107 mg

Just Like Tuna Salad

MAKES 4 CUPS (1 L)

2 cups	cooked pinto beans or 1 can (19 oz [540 mL]), rinsed and drained	500 mL
3 1/2 cups	coarsely chopped artichoke hearts	875 mL
3	green onions, chopped	3
2 tbsp	lemon juice	25 mL
2 tbsp	chopped parsley	25 mL
2 tsp	olive oil	10 mL
2	cloves garlic, minced	2
1/2 tsp	salt	2 mL

1. In a food processor, combine beans, 1 1/2 cups (375 mL) of the artichoke hearts, green onions, lemon juice, parsley, olive oil, garlic and salt; process until smooth.

2. Transfer mixture to a bowl and add remaining artichoke hearts. With a fork, stir lightly until mixture achieves a flaky ("tuna-like") consistency. Refrigerate before serving.

The artichokes in this recipe create a flaky texture similar to tuna, while the beans provide the calcium. Our guests are always pleasantly surprised by this tasty and unique tuna substitute which can be spread over any type of bread or crackers. When serving, place in a decorative bowl and garnish with sliced cherry tomatoes or chopped green onions.

CALCIUM NEWS

Although conventional food guides call for 2 to 3 servings of dairy products per day, many population groups are genetically unaccustomed to digesting dairy proteins and obtain their calcium from plant-based foods.

≺ CHUNKY CORN CHOWDER (PAGE 48)

PER 1/2 CUP (125 mL)								
Calcium	Calories	Protein	Fat	Saturated Fat	Carbohydrates	Fiber	Iron	Sodium
61 mg	109	6 g	2 g	0.2 g	20 g	8 g	2 mg	216 mg

Believe It Or Not Egg Salad

Makes 2 cups (500 mL)

4 oz	soft tofu	125 g
8 oz	extra firm tofu, crumbled	250 g
2	green onions, finely chopped	2
2	small celery stalks, finely chopped	2
1 tbsp	chopped parsley	15 mL
3/4 tsp	garlic salt	4 mL
1/4 tsp	paprika	1 mL
1/4 tsp	turmeric	1 mL
1/4 tsp	salt	1 mL
1/8 tsp	curry powder	0.5 mL
1/8 tsp	black pepper	0.5 mL

1. In a blender or food processor, purée soft tofu. Transfer to a bowl. Add crumbled tofu and mix in. Add green onions, celery and parsley.

2. In another bowl, combine remaining ingredients. Add to tofu mixture. Gently toss together until well mixed. Refrigerate 1 hour and serve cold.

*T*IP

The color of this "egg" salad is light yellow when it is first made; however, if you let it sit in the refrigerator for a few hours or overnight, it will turn a deeper "egg" yellow color.

PER 1/2 CUP (125 mL)

Calcium	Calories	Protein	Fat	Saturated Fat	Carbohydrates	Fiber	Iron	Sodium
195 mg	81	11 g	2 g	0.2 g	6 g	3 g	0.4 mg	394 mg

Mock Mayonnaise

MAKES 1 1/4 CUPS (300 ML)

5 oz	*firm tofu*	150 g
3 oz	*regular tofu* or *soft tofu*	75 g
2 tbsp	*lemon juice*	25 mL
1 1/2 tbsp	*sugar*	20 mL
1	*clove garlic, minced*	1
1/2 tbsp	*almond butter*	7 mL
1/2 tsp	*salt*	2 mL

1. In a blender or food processor, blend all ingredients until smooth. Serve immediately or refrigerate and serve cold.

*T*his versatile calcium-charged mayonnaise makes a light-tasting dressing with a nice tang. It's wonderful for spreading on sandwiches or as a dressing for potato salad — or simply as a dip for raw vegetables.

You can find almond butter in most health food stores.

CALCIUM FACT

Tofu is not only an excellent source of calcium, but is also believed to lower the risk of heart disease and some forms of cancer.

PER 2 TBSP (25 mL)								
Calcium	Calories	Protein	Fat	Saturated Fat	Carbohydrates	Fiber	Iron	Sodium
41 mg	40	3 g	2 g	0.3 g	3 g	0.2 g	2 mg	118 mg

Sassy Garlic Dip

MAKES 1 1/2 CUPS

If you're in a hurry to put together a dip for those unexpected guests, then this is the one for you! Easy to make with a bit of a garlicky bite — this dip is one of our favorites.

8 oz	soft tofu	250 g
2 tbsp	lemon juice	25 mL
3	cloves garlic, minced	3
1/2 cup	white kidney beans, cooked	125 mL
1/4 tsp	salt	1 mL
1	green onion, chopped	1

1. In a food processor or blender, combine tofu, lemon juice, garlic, beans and salt. Process until smooth. Add green onion and process a few seconds, trying not to liquefy the onion. Refrigerate and serve cold or at room temperature.

*T*IP

Canned beans are widely available and are one of the best time-savers when preparing your calcium-rich recipes.

PER 2 TBSP (25 mL)

Calcium	Calories	Protein	Fat	Saturated Fat	Carbohydrates	Fiber	Iron	Sodium
37 mg	24	2 g	1 g	0.1 g	3 g	1 g	0 mg	52 mg

Soups

Soup has long been a welcome mealtime staple — perfect for lunch or dinner. Served as an entrée or with bread as a complete meal, these soup recipes will satisfy the heartiest appetites and most discerning palates. Even better, they pack a real calcium punch.

To add body to a soup, try adding small pasta shells during the last 10 to 15 minutes of cooking time. To boost the calcium, simply add some cooked quinoa, beans, or 1/2-inch (1 cm) cubes of tofu in the final stages of cooking.

To increase the calcium even further, shred some greens such as kale, collards, beet greens or other high-calcium greens and toss them into your soup. You can also add shredded parsnips, chopped green beans or any number of different vegetables. (Check the Calcium Table on page 174.)

Soups keep well in the freezer, so don't hesitate to store leftovers. Just remember to leave a little space between the soup and lid, since liquids expand when frozen.

Green Bean Parsnip Soup

SERVES 4 TO 6

Light, peppery and slightly chunky, this soup goes well with any main course. A very attractive dish when garnished with a sprinkling of finely shredded carrots at the center.

1 tbsp	olive oil	15 mL
1/4 cup	chopped onions	50 mL
5 cups	vegetable stock	1.25 L
2 cups	chopped green beans	500 mL
2 cups	finely chopped parsnips	500 mL
2 cups	packed chopped bok choy	500 mL
1 cup	chopped peeled potatoes	250 mL
1 tsp	dried basil	5 mL
1 tsp	dried thyme	5 mL
1/4 tsp	black pepper	1 mL

1. In a large pot with a lid, heat oil over medium-high heat. Add onions; cook 2 minutes or until softened.

2. Add remaining ingredients and, stirring frequently, bring to a boil. Reduce heat, cover and simmer for 10 minutes.

3. In batches, using a food processor or blender, process soup to a chunky consistency. Return soup to pot. Heat and serve.

PER SERVING (4)								
Calcium	Calories	Protein	Fat	Saturated Fat	Carbohydrates	Fiber	Iron	Sodium
119 mg	166	4 g	5 g	0.6 g	29 g	5 g	2 mg	811 mg

Potato Soup with Greens

SERVES 4 TO 6

1 tbsp	olive oil	15 mL
3 cups	packed chopped kale or turnip greens	750 mL
2 cups	packed chopped beet greens	500 mL
5 1/2 cups	vegetable stock	1.375 L
2	medium white potatoes, peeled and chopped into small pieces	2
2	sweet potatoes, peeled and chopped into small pieces	2
1/4 cup	lemon juice	50 mL
1 tbsp	soy sauce	15 mL

1. In a nonstick saucepan, heat oil over medium heat. Add kale and beet greens; sauté 5 minutes. Set aside.

2. In a large pot, bring vegetable stock to a boil. Add potatoes and simmer, covered, for 20 minutes.

3. Transfer 3 cups (750 mL) of soup mixture (liquids and solids) to food processor or blender; process until smooth. Return to pot. Add greens, lemon juice and soy sauce; simmer, covered, for another 15 minutes before serving.

This tangy soup has an appealing orange broth that doesn't need additional vegetables to add interest or flavor. We frequently have this delicious soup as our main course with a variety of hearty breads — and we always indulge in a second or third helping.

TIP

Beet greens are mildly flavored and have a soft consistency when cooked. Feel free to add these greens to soups to boost the calcium.

PER SERVING (4)								
Calcium	Calories	Protein	Fat	Saturated Fat	Carbohydrates	Fiber	Iron	Sodium
196 mg	310	8 g	5 g	0.6 g	62 g	8 g	3 mg	1262 mg

Comfy Kale Soup

Serves 6

Deliciously green and creamy, this soup is packed full of calcium. It's mild, yet flavorful. And since this soup is made with loads of green leafy kale, you may find there is no need to serve a salad.

1 1/2 tbsp	olive oil	20 mL
3	large cloves garlic, chopped	3
3	medium leeks, sliced	3
1	parsnip, chopped	1
1	medium red onion, chopped	1
1	stalk celery, thinly sliced	1
6 cups	vegetable stock	1.5 L
1	bay leaf	1
1 tsp	dried parsley	5 mL
1 tsp	dried tarragon	5 mL
1/4 tsp	white pepper	1 mL
8 cups	packed chopped kale	2 L
1 cup	soy milk (calcium fortified)	250 mL

1. In a large pot with a lid, heat oil over medium-high heat. Add garlic, leeks, parsnip, onion and celery; sauté for 5 minutes.

2. Add vegetable stock, bay leaf, parsley, tarragon, white pepper and kale. Mix together and bring to a boil. Reduce heat, cover and simmer for 15 minutes.

3. Remove bay leaf — if you can find it! In batches, with a food processor or blender, purée vegetable mixture. Return puréed soup to pot and add soy milk. Heat and serve.

TIPS

For a clear broth soup, simply skip Step 3. Remember to remove the bay leaf, however.

Other high-calcium greens can replace the kale in this soup. Try collards, beet greens or mustard greens.

		PER SERVING						
Calcium	Calories	Protein	Fat	Saturated Fat	Carbohydrates	Fiber	Iron	Sodium
272 mg	175	7 g	6 g	0.7 g	28 g	6 g	4 mg	710 mg

Chunky Brussels Sprouts Soup

SERVES 4

1 tbsp	olive oil	15 mL
2	parsnips, finely chopped	2
2	carrots, chopped	2
1	onion, chopped	1
2 cups	fresh Brussels sprouts, cut into quarters	500 mL
1/2 cup	sliced mushrooms	125 mL
3 cups	vegetable stock	750 mL
1/4 tsp	black pepper	1 mL
1/8 tsp	nutmeg	0.5 mL
1 cup	soy milk (calcium fortified)	250 mL

Here's an interesting and tasty way to serve Brussels sprouts to your family or guests. The parsnips in this chunky soup create a light semi-sweet flavor.

1. In a large pot with a lid, heat oil over medium-high heat. Add parsnips, carrots and onion; sauté approximately 3 minutes. Add Brussels sprouts and mushrooms. Cook, stirring, for an additional 5 minutes. Stir in vegetable stock and bring to a boil. Reduce heat, cover and simmer for 15 minutes.

2. Transfer approximately half the mixture to a food processor or blender; purée until smooth. Return puréed soup to pot. Stir to mix.

3. Over low heat, add pepper and nutmeg to pot. Slowly stir in soy milk; simmer for 5 minutes, then serve.

TIP

If fresh Brussels sprouts are unavailable, use frozen. The soup will be just as tasty.

	PER SERVING							
Calcium	Calories	Protein	Fat	Saturated Fat	Carbohydrates	Fiber	Iron	Sodium
139 mg	166	5 g	6 g	0.7 g	27 g	7 g	2 mg	524 mg

Carrot Broccoli Soup with Fresh Dill

SERVES 6

We love this rich, brothy soup because it's loaded with our favorite calcium-rich bean — soybeans — and it has a tasty and prominent dill flavor. The broccoli and chunks of tofu not only add texture but they boost the calcium as well.

1 tbsp	olive oil	15 mL
1	medium onion, chopped	1
3 cups	cooked soybeans or 2 cans (each 14 oz [425 g]), rinsed and drained	750 mL
7 cups	vegetable stock	1.75 L
6 cups	broccoli florets, cut into bite-size pieces	1.5 L
3	carrots, chopped	3
2 tbsp	chopped dill	25 mL
1 cup	soy milk (calcium fortified)	250 mL
8 oz	soft tofu, cut into 1/4-inch (5 mm) cubes	250 g

1. In a large pot with a lid, heat oil over medium-high heat. Add onions and cook until golden. Add soybeans, vegetable stock, broccoli, carrots and dill. Bring to a boil. Reduce heat and simmer, covered, for 25 minutes.

2. Add soy milk and tofu, stirring gently, and continue to simmer for an additional 5 minutes. Serve hot.

CALCIUM QUOTE

"Diets rich in protein, especially animal protein, are known to cause people to excrete more calcium ... and increase their risk of osteoporosis."
Physicians' Committee for Responsible Medicine

				PER SERVING				
Calcium	Calories	Protein	Fat	Saturated Fat	Carbohydrates	Fiber	Iron	Sodium
261 mg	274	22 g	13 g	1.7 g	23 g	9 g	6 mg	795 mg

Very Veggie Minestrone

SERVES 6

1 tbsp	olive oil	15 mL
1/2 cup	finely chopped onions	125 mL
1/2 cup	chopped celery	125 mL
2 cups	cooked white, black or pinto beans or 1 can (19 oz [540 mL]), rinsed and drained	500 mL
4 cups	vegetable stock	1 L
3 cups	chopped fresh tomatoes	750 mL
2 cups	packed finely chopped beet greens	500 mL
1 cup	halved green beans	250 mL
1/2 tsp	dried ground rosemary	2 mL
1 cup	chopped fresh okra	250 mL
1/4 tsp	salt	1 mL
	Black pepper, to taste	

1. In a large pot with a lid, heat oil over medium-high heat. Add onions and celery; sauté for 5 minutes.

2. Stir in beans, stock, tomatoes, beet greens, green beans and rosemary. Bring to a boil. Reduce heat, cover and simmer 15 minutes, stirring occasionally.

3. Add okra, salt and pepper; simmer, covered, an additional 10 minutes before serving.

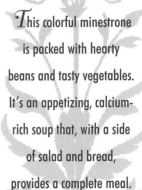

This colorful minestrone is packed with hearty beans and tasty vegetables. It's an appetizing, calcium-rich soup that, with a side of salad and bread, provides a complete meal.

TIPS

For a thicker consistency, small pasta shells can be added in Step 2. Use 1/2 cup to 3/4 cup (125 to 175 mL) pasta.

Beet greens, turnip greens, kale and collards are so versatile that they can be used interchangeably in most soup recipes.

				PER SERVING				
Calcium	Calories	Protein	Fat	Saturated Fat	Carbohydrates	Fiber	Iron	Sodium
117 mg	156	8 g	3 g	0.4 g	26 g	7 g	4 mg	591 mg

Loads of Lentils Soup

Serves 8

2 tsp	canola oil	10 mL
2	onions, halved and sliced	2
4	cloves garlic, minced	4
10 cups	water	2.5 L
2 cups	red lentils, rinsed and drained	500 mL
1	can (5 1/2 oz [156 mL]) tomato paste	1
3	parsnips, peeled and sliced	3
1 tbsp	salt	15 mL
3 tbsp	chopped parsley, divided	45 mL
1 tsp	dried oregano	5 mL
1 tsp	dried basil	5 mL
1 tsp	black pepper	5 mL
5 cups	packed chopped dandelion greens or kale	1.25 L

1. In a large pot with a lid, heat oil over medium-high heat. Add onions and sauté for 3 minutes. Add garlic and sauté 1 minute more.

2. Add water, lentils, tomato paste, parsnips, salt, 1 tbsp (15 mL) parsley, oregano, basil and pepper; bring to a boil. Reduce heat to low, cover and simmer for 30 minutes. Add dandelion greens and simmer an additional 15 minutes or until greens are soft. Serve hot soup in bowls garnished with remaining parsley.

CALCIUM FACT

Calcium research suggests that calcium in green leafy vegetables is absorbed by the body more efficiently than cow's milk.

PER SERVING

Calcium	Calories	Protein	Fat	Saturated Fat	Carbohydrates	Fiber	Iron	Sodium
137 mg	261	15 g	2 g	0.2 g	49 g	10 g	7 mg	924 mg

Yellow Bean Curry Soup

SERVES 6 TO 8

1 tbsp	olive oil	15 mL
2	large onions, chopped	2
4	cloves garlic, minced	4
3 cups	packed chopped Chinese lettuce	750 mL
2 tsp	curry powder	10 mL
6 cups	water	1.5 L
3 cups	chopped yellow beans	750 mL
6 oz	soft tofu, crumbled	175 g
3/4 cup	tomato sauce	175 mL
2	red bell peppers, seeded and chopped	2
2 tbsp	lemon juice	25 mL
1 1/2 tbsp	soy sauce	20 mL
1	bay leaf	1
1 tbsp	chopped parsley	15 mL
2 tsp	salt	10 mL
1/2 tsp	dried thyme	2 mL

1. In a large pot, heat oil over medium heat. Add onions and garlic; cook until soft and golden.

2. Add Chinese lettuce and curry powder; sauté approximately 2 minutes. Add remaining ingredients and bring to a boil. Reduce heat, cover and simmer for 20 minutes.

3. Remove bay leaf and serve soup hot.

With a hint of curry and an abundance of colorful vegetables, this tasty soup is an enticing delight to serve. This soup is perfect for fortifying with additional calcium-rich ingredients because it has an abundance of broth. Toss in some chopped beet greens or kale during the last 5 minutes of cooking or add small squares of soft or regular tofu. (We often do this with many of our soups.)

		PER SERVING (6)						
Calcium	Calories	Protein	Fat	Saturated Fat	Carbohydrates	Fiber	Iron	Sodium
142 mg	113	5 g	4 g	0.5 g	18 g	4 g	2 mg	1230 mg

Black Bean Dream Soup

SERVES 6 TO 8

Rich, dark and creamy, this soup is a wonderfully tasty treat for those damp and chilly days. With a side of hearty bread, it can easily become a complete meal. For an attractive garnish, place fresh mint leaves at the center of each bowl.

2 cups	dried black beans	500 mL
5 cups	water	1.25 L
2 cups	vegetable stock	500 mL
4 cups	packed chopped bok choy	1 L
1	onion, chopped	1
4	cloves garlic, minced	4
1	large sweet potato, peeled and chopped	1
1	bay leaf	1
1/2 cup	tomato sauce	125 mL
2 tsp	salt	10 mL
1/2 tsp	dried thyme	2 mL
1/2 tsp	black pepper	2 mL

1. Soak beans in water for 3 hours or overnight. Drain, discarding soaking liquid.

2. In a large pot with a lid, combine beans, vegetable stock, bok choy, onion, garlic, sweet potato and bay leaf. Bring to a boil, reduce heat and simmer, covered, for 1 1/2 hours.

3. Remove bay leaf. In small batches, with a food processor or blender, purée soup until smooth and creamy, returning puréed soup to the pot.

4. Stir in tomato sauce, salt, thyme and pepper. Simmer, covered, over low heat for 10 minutes. Serve hot.

PER SERVING (6)

Calcium	Calories	Protein	Fat	Saturated Fat	Carbohydrates	Fiber	Iron	Sodium
185 mg	305	16 g	1 g	0.2 g	61 g	14 g	6 mg	1138 mg

Chickpea Lemon Soup

SERVES 4

3 cups	cooked chickpeas or 2 cans (each 14 oz [398 mL]), rinsed and drained	750 mL
3 cups	vegetable stock	750 mL
1	red bell pepper, chopped	1
3	small onions, chopped	3
4	large cloves garlic, minced	4
1	bay leaf	1
3 tbsp	lemon juice	45 mL
2 tbsp	chopped parsley	25 mL
2 tsp	olive oil	10 mL
1/2 tsp	salt	2 mL
1/8 tsp	black pepper	0.5 mL

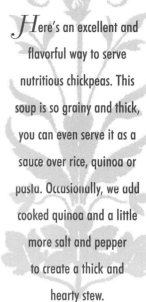

Here's an excellent and flavorful way to serve nutritious chickpeas. This soup is so grainy and thick, you can even serve it as a sauce over rice, quinoa or pasta. Occasionally, we add cooked quinoa and a little more salt and pepper to create a thick and hearty stew.

1. In a large pot with a lid, combine all ingredients, cover and bring to a boil. Reduce heat and simmer for 20 minutes.

2. Remove bay leaf. In batches, with a blender or food processor, purée soup until smooth, returning puréed soup to pot. Heat for another 5 minutes and serve.

TIP

Try garnishing this soup with a sprinkling of toasted bread crumbs at the center of each bowl.

				PER SERVING				
Calcium	Calories	Protein	Fat	Saturated Fat	Carbohydrates	Fiber	Iron	Sodium
86 mg	270	12 g	6 g	0.7 g	44 g	6 g	4 mg	776 mg

Chunky Corn Chowder

SERVES 6

Our entire family loves this hearty and flavorful corn chowder. It's a colorful soup creation and, best of all, full of calcium-rich vegetables!

4 cups	*chopped peeled sweet potatoes*	*1 L*
4 cups	*vegetable stock*	*1 L*
2	*small onions, chopped*	*2*
2 cups	*corn kernels, drained*	*500 mL*
2 cups	*chopped packed Chinese lettuce*	*500 mL*
3	*stalks celery, chopped*	*3*
3	*tomatoes, peeled and chopped*	*3*
1 tsp	*salt*	*5 mL*
1 cup	*okra, cut into rounds 1/4 inch (5 mm) thick*	*250 mL*
1 cup	*packed finely chopped collard greens*	*250 mL*
1 cup	*soy milk (calcium fortified)*	*250 mL*
	Chopped parsley or cilantro	

1. In a large pot with a lid, combine sweet potatoes, stock, onions, corn, Chinese lettuce, celery, tomatoes and salt. Cover and bring to a boil. Reduce heat and simmer for 20 minutes.

2. Add okra and collard greens; simmer for an additional 20 minutes.

3. Gradually stir in soy milk. Serve garnished with a small amount of chopped parsley.

PER SERVING

Calcium	Calories	Protein	Fat	Saturated Fat	Carbohydrates	Fiber	Iron	Sodium
151 mg	199	6 g	2 g	0.2 g	43 g	7 g	2 mg	849 mg

Butternut Squash Soup

SERVES 6

1	large butternut squash	1
1 tbsp	olive oil	15 mL
1	small onion, chopped	1
1	large leek, white part only, thinly sliced	1
4 1/2 cups	vegetable stock	1.125 L
2 cups	finely chopped broccoli	500 mL
1 tsp	salt	5 mL
1/2 tsp	dried thyme	2 mL
1/4 tsp	dried sage	1 mL
1/8 tsp	ground nutmeg	0.5 mL
4 cups	packed finely chopped turnip greens or collard greens, steamed	500 mL
1 cup	soy milk (calcium fortified)	250 mL
	Black pepper to taste	
	Chopped chives	

We love this combination of butternut squash and soy milk, as it becomes a creamy-rich broth when blended. Add in the broccoli and greens and you will have a flavorful, smooth, calcium-rich soup.

1. Cut squash in half and remove seeds. Peel and chop squash into 2-inch (5 cm) chunks.

2. In a large pot with a lid, heat oil over medium heat. Add onion and leek; sauté for 3 minutes.

3. Stir in stock, squash, broccoli, salt, thyme, sage and nutmeg; bring to a boil. Reduce heat, cover and simmer 30 minutes.

4. In batches, using a food processor or blender, purée soup. Return soup to pot. Stir in turnip greens and soy milk; simmer 3 minutes. Serve garnished with chopped chives.

PER SERVING

Calcium	Calories	Protein	Fat	Saturated Fat	Carbohydrates	Fiber	Iron	Sodium
314 mg	191	6 g	4 g	0.5 g	38 g	8 g	3 mg	909 mg

Cool and Minty Bean Soup

SERVES 4 TO 6

2 cups	chopped peeled carrots	500 mL
1/2 cup	chopped celery	125 mL
2 cups	cooked red kidney beans or 1 can (19 oz [540 mL]), rinsed and drained	500 mL
1 cup	diced seedless cucumber	250 mL
3 cups	tomato juice, divided	750 mL
1/2 cup	water	125 mL
1/4 cup	lime juice	50 mL
3	cloves garlic, minced	3
1/4 cup	finely chopped mint	50 mL

1. Steam carrots and celery together until tender. Transfer to large bowl and set aside to cool.

2. Add beans, cucumber, 2 cups (500 mL) tomato juice, water, lime juice and garlic to bowl. Transfer mixture in batches to food processor or blender and process until well blended.

3. Transfer soup to another large bowl and add remaining tomato juice and mint. Mix well. Refrigerate 3 hours to blend flavors. Serve cold.

Wonderfully cool and refreshing, this minty, tomato-based soup is the perfect dish to serve on those hot summer days.

TIP

To give this soup a spicy kick, simply add 1/4 tsp (1 mL) chili flakes in Step 1. Or, to add color, you can add 4 oz (125 mg) soft tofu and 1 small avocado diced into small cubes.

PER SERVING (4)								
Calcium	Calories	Protein	Fat	Saturated Fat	Carbohydrates	Fiber	Iron	Sodium
96 mg	181	10 g	1 g	0.1 g	37 g	9 g	4 mg	696 mg

Salads & Salad Dressings

One of the best parts of any meal — and our personal favorite — is the salad, a wholesome, refreshing selection of fresh greens, beans, pasta or other natural ingredients. The recipes in this section will let you create a complete meal that offers loads of flavor without compromising nutrients, particularly calcium.

Try these salads and dressings as the focus of your next meal. Each calcium-rich salad will awaken your senses with a wonderful array of color and texture. Each of these creations — be it a Waldorf Salad with a twist, a novel Chickpea Spiral Pasta Salad or a zesty Cilantro Bean Salad — will be a pleasure to serve your dinner companions.

Two-Bean Salad

SERVES 4

4	small plum tomatoes, seeded and diced	4
1	small seedless cucumber, peeled and diced	1
1 1/2 cups	cooked Great Northern beans or 1 can (14 oz [398 mL]), rinsed and drained	375 mL
1 1/2 cups	cooked red kidney beans or 1 can (14 oz [398 mL]), rinsed and drained	375 mL
1/4 cup	chopped parsley	50 mL
3 tbsp	white wine vinegar	45 mL
1 1/2 tbsp	olive oil	20 mL
2	cloves garlic, minced	2
1/2 tsp	salt	2 mL
1/8 tsp	black pepper	0.5 mL
	Mint leaves	

CALCIUM FACT

Negative calcium balance is the process of human metabolism where more calcium is lost through the urine than is absorbed from the diet. Excess salt and animal protein are thought to be major contributors to negative calcium balance.

1. In a large bowl, combine tomatoes, cucumber, beans and parsley.

2. In a small bowl, whisk together vinegar, oil, garlic, salt and pepper. Pour dressing over salad and toss. Refrigerate and serve cold, garnished with mint leaves.

PER SERVING

Calcium	Calories	Protein	Fat	Saturated Fat	Carbohydrates	Fiber	Iron	Sodium
93 mg	203	12 g	3 g	0.5 g	33 g	8 g	4 mg	296 mg

Cilantro Bean Salad

Serves 4

Preheat oven to 400° F (200° C)

1	small butternut squash	1
3	parsnips, sliced	3
1/2 tbsp	olive oil	7 mL
2 cups	chopped onions	500 mL
3	large cloves garlic, minced	3
2 tsp	ground cumin	10 mL
1 1/2 tsp	dried tarragon	7 mL
4 cups	cooked white or black beans or *2 cans (each 19 oz [540 mL]), rinsed and drained*	1 L
1/2 cup	chopped cilantro	125 mL
3/4 cup	vegetable stock	175 mL
3 tbsp	lemon juice	45 mL
1/2 tsp	salt	2 mL

1. Cut squash in half; remove and discard seeds. Wrap in foil and bake for 1 1/2 hours or until soft. Scoop flesh out into a bowl (you should get approximately 2 1/2 cups [625 mL]) and mash. Set aside.

2. Steam or boil parsnips until slightly softened. Set aside.

3. In a large nonstick skillet, heat oil over medium-high heat. Add onions, garlic, cumin and tarragon; cook for 3 to 4 minutes or until softened. (Add 1 tsp [5 mL] water if onions become dry.) Add cooked squash, beans, cilantro, stock, lemon juice and salt. Reduce heat to low and cook for 10 minutes, stirring often. Remove from heat. Serve immediately or refrigerate and serve cold.

One of our very favorites, this thick and creamy bean salad has the wonderful aroma and flavor of cilantro (fresh coriander). Serve hot or cold, scooped over a bed of lettuce and garnished with cilantro leaves.

TIP

For an interesting presentation, try using this salad as a stuffing for tomatoes.

PER SERVING								
Calcium	Calories	Protein	Fat	Saturated Fat	Carbohydrates	Fiber	Iron	Sodium
301 mg	456	22 g	3 g	0.5 g	92 g	19 g	9 mg	433 mg

Nutty Waldorf Salad

SERVES 6 TO 8

This sweet and crunchy salad is perfect as a side dish or, with the addition of hearty bread slices, a complete meal.

6 oz	regular or firm tofu, crumbled	185 g
2	medium red apples, cored and chopped	2
1	medium tart green apple, cored and chopped	1
2	seedless oranges, peeled and cut into bite-size pieces	2
1	stalk celery, finely chopped	1
1/2 cup	chopped dried figs	125 mL

Dressing

1/4 cup	soy yogurt	50 mL
1/2 tsp	grated lemon zest	2 mL
Half	small lemon (juice only)	Half
Half	ripe avocado, mashed	Half
2 1/2 tbsp	granulated sugar	35 mL
1 1/2 tbsp	maple syrup	20 mL
3/4 cup	toasted slivered almonds	175 mL
1/2 cup	chopped walnuts	125 mL

TIP

To toast almonds: Place almonds on baking sheet. Heat under broiler for 2 to 3 minutes or until lightly browned (stir after the first minute).

1. In a large bowl, combine tofu, apples, oranges, celery and figs. Set aside.

2. Make the dressing: In a blender or food processor, combine yogurt, lemon zest and juice, avocado, sugar and maple syrup; purée until smooth.

3. Pour dressing over salad and toss. Serve at room temperature or refrigerate if you enjoy your salads cold. Sprinkle almonds and walnuts over salad and mix together before serving.

PER SERVING (6)								
Calcium	Calories	Protein	Fat	Saturated Fat	Carbohydrates	Fiber	Iron	Sodium
129 mg	347	9 g	19 g	2.1 g	42 g	6 g	3 mg	19 mg

Carrot Broccoli Salad with Cilantro

SERVES 6

4 cups	chopped broccoli florets	1 L
2 cups	finely chopped bok choy	500 mL
3	carrots, peeled and coarsely grated	3
3 tbsp	dark sesame oil, divided	45 mL
2	cloves garlic, minced	2
12 oz	firm tofu, cut into 1/4-inch (5 mm) cubes	375 g
1/4 cup	sesame seeds	50 mL
1/2 cup	chopped cilantro	125 mL

Dressing

1/4 cup	lemon juice	50 mL
1/2 tsp	salt	2 mL
1/4 tsp	black pepper	1 mL

If you enjoy the taste of sesame, this calcium-rich salad is the perfect choice. We frequently make this colorful, crunchy, refreshing salad because it complements any meal.

1. Lightly steam broccoli until tender-crisp, about 3 minutes. Add bok choy and carrots; steam for another 2 minutes. Transfer vegetables to a large bowl and set aside.

2. In a nonstick skillet, heat 1 tbsp (15 mL) sesame oil over medium-high heat. Add garlic and tofu; cook for 3 minutes. Add sesame seeds; cook for 3 minutes more. Scrape mixture over steamed vegetables. Add cilantro; mix well.

3. In a small bowl, whisk together lemon juice, remaining 2 tbsp (25 mL) sesame oil, salt and pepper. Pour dressing over salad and toss. Refrigerate 1 hour and serve cold.

TIPS

This salad is equally delicious whether served at room temperature or chilled.

Try combining this salad with additional high-calcium greens to provide even more calcium.

PER SERVING

Calcium	Calories	Protein	Fat	Saturated Fat	Carbohydrates	Fiber	Iron	Sodium
189 mg	198	13 g	13 g	1.9 g	11 g	4 g	7 mg	162 mg

Kale Pasta Salad

SERVES 4 TO 6

7 oz	rotini	200 g
6 cups	packed finely chopped kale, ribs removed	1.5 L
2 1/2 cups	cooked white or black beans or 1 can (19 oz [540 mL]), rinsed and drained	625 mL
8 oz	firm tofu, crumbled	250 g
2	plum tomatoes, seeded and diced	2
1/4 cup	chopped pitted green olives	50 mL
2 tbsp	capers	25 mL

Dressing

1/3 cup	red wine vinegar	75 mL
1 1/2 tbsp	olive oil	20 mL
3	cloves garlic, minced	3
1 tsp	salt	5 mL
1/4 tsp	black pepper	1 mL

1. In a large pot of boiling water, cook pasta until tender but firm; drain. Rinse with cold water and drain. Set aside.

2. Steam kale until wilted; let cool. With your hands, squeeze out excess water. Set aside.

3. In a large bowl, combine pasta, kale, beans, tofu, tomatoes, olives and capers.

4. In a small bowl, whisk together vinegar, oil, garlic, salt and pepper. Pour dressing over salad and toss to coat well.

PER SERVING (4)								
Calcium	Calories	Protein	Fat	Saturated Fat	Carbohydrates	Fiber	Iron	Sodium
374 mg	539	30 g	13 g	1.9 g	81 g	13 g	13 mg	887 mg

Chickpea Spiral Pasta Salad

SERVES 6

6 oz	spiral pasta	175 g
3 cups	cooked Great Northern or white beans or 2 cans (each 14 oz [398 mL]), rinsed and drained	750 mL
1 cup	cooked chickpeas or 1 can (14 oz [398 mL]), rinsed and drained	250 mL
3/4 cup	chopped drained sun-dried tomatoes	175 mL
1/2 cup	finely chopped red bell peppers	125 mL

Dressing

1	lemon (juice only)	1
1 tbsp	olive oil	15 mL
1/4 cup	chopped dill	50 mL
1/4 cup	tofu mayonnaise	50 mL
3	large cloves garlic, minced	3
1 1/2 tsp	garlic salt	7 mL
3/4 cup	toasted slivered almonds (for technique see Tip, page 54)	175 mL

The combination of sun-dried tomatoes, lemon and garlic turns this salad into a delicious favorite. To make this dish more interesting, use colored spiral pasta to brighten it up.

1. In a large pot of boiling water, cook pasta until tender but firm; drain. Rinse with cold water and drain.

2. In a large bowl, combine beans and chickpeas. Add pasta and toss. Add tomatoes and red peppers; toss.

3. In a small bowl, whisk together lemon juice, oil, dill, mayonnaise, garlic and garlic salt. Pour dressing over salad, and sprinkle with almonds. Toss and serve.

PER SERVING

Calcium	Calories	Protein	Fat	Saturated Fat	Carbohydrates	Fiber	Iron	Sodium
150 mg	420	19 g	15 g	1.7 g	57 g	7 g	5 mg	380 mg

Creamy Coleslaw

SERVES 4

1 cup	broccoli florets	250 mL
3	small carrots	3
1	small onion, peeled and quartered	1
4	large leaves Chinese lettuce	4

Dressing

4 oz	soft tofu	125 g
2 tbsp	tofu mayonnaise	25 mL
1 1/2 tbsp	granulated sugar	20 mL
1 tbsp	red wine vinegar	15 mL
1 tbsp	maple syrup	15 mL
1 tbsp	lemon juice	15 mL
1/8 tsp	celery salt	0.5 mL
2 tsp	sesame seeds	10 mL

1. Using a food processor with a grater attachment, separately shred the broccoli, carrots, onion and Chinese lettuce. After grating the Chinese lettuce, squeeze out the excess liquid. Combine shredded vegetables in a bowl and toss together.

2. In a blender or food processor, combine tofu, mayonnaise, sugar, vinegar, maple syrup, lemon juice and celery salt. Process until a creamy consistency is achieved.

3. Pour dressing over vegetables and toss. Sprinkle with sesame seeds and toss. Serve immediately or refrigerate for 1 hour and serve cold.

Your guests and family will marvel at this delicious combination of creamy dressing and shredded high-calcium veggies. It's a colorful and zesty variation on a familiar dish.

CALCIUM QUOTE

"Plant foods, naturally low in fat and high in nutrition, are known to protect against many cancers."

Consumer Reports on Health, July 1997

PER SERVING								
Calcium	Calories	Protein	Fat	Saturated Fat	Carbohydrates	Fiber	Iron	Sodium
120 mg	105	5 g	3 g	0.4 g	18 g	2 g	1 mg	98 mg

Perfect Potato Salad

SERVES 4

4	large potatoes, peeled and quartered	4
8 oz	firm tofu, cut into 1/4-inch (5 mm) cubes	250 g
3	green onions, chopped	3
1	stalk celery, finely chopped	1
1	small red bell pepper, finely chopped	1
2 tbsp	chopped dill	25 mL
1 tsp	salt	5 mL
1 cup	MOCK MAYONNAISE (see recipe page 35) or store-bought variety	250 mL

1. In a large pot of boiling water, cook potatoes until tender but firm; drain. Refresh under cold water and drain. Chop into 1-inch (2.5 cm) cubes. Place in a large bowl and set aside.

2. Add remaining ingredients to bowl and toss together. Serve immediately or refrigerate and serve cold.

Our family loves this creamy potato salad because it is light yet hearty. The celery and red pepper add a nice color and crunch, with the mock mayo and tofu providing additional calcium. You'll find this is a wonderful alternative to traditional potato salad.

	PER SERVING							
Calcium	Calories	Protein	Fat	Saturated Fat	Carbohydrates	Fiber	Iron	Sodium
294 mg	361	21 g	11 g	1.5 g	52 g	6 g	13 mg	907 mg

Yellow and Green Bean Salad with Artichokes

SERVES 4

This delightfully succulent salad is a colorful presentation of green and yellow beans. The tofu absorbs the tasty marinade and gives this dish its calcium boost.

1/2 cup	vegetable stock	125 mL
2 tbsp	chopped cilantro	25 mL
1 tbsp	soy sauce	15 mL
2 tsp	prepared mustard	10 mL
1 tbsp	lemon juice	15 mL
8 oz	firm tofu, cut into 1/4-inch (5 mm) chunks	250 g
1 cup	chopped artichoke hearts	250 mL
1 cup	halved green beans	250 mL
1 cup	halved yellow beans	250 mL
1	large red bell pepper, diced	1
3/4 cup	corn kernels, drained	175 mL

1. In a bowl whisk together vegetable stock, cilantro, soy sauce, mustard and lemon juice. Add tofu and artichoke hearts; toss to coat. Refrigerate, covered, to marinate for 1 hour.

2. Steam or boil green and yellow beans until tender. Let cool.

3. Transfer beans to tofu-artichoke mixture. Add red pepper and corn; toss together until well mixed. Serve cold.

TIP

If you don't care for the taste of cilantro, you can substitute chopped fresh mint or rosemary.

PER SERVING

Calcium	Calories	Protein	Fat	Saturated Fat	Carbohydrates	Fiber	Iron	Sodium
169 mg	168	13 g	6 g	0.8 g	22 g	5 g	8 mg	428 mg

Wild Rice Quinoa Salad

SERVES 6

2 cups	cooked quinoa (see page 20)	500 mL
1 cup	cooked wild rice	250 mL
2	tomatoes, seeded and diced	2
3/4 cup	chopped parsley	175 mL
1 cup	coarsely chopped dried figs	250 mL
1/2 cup	chopped green onions	125 mL

Dressing

1/2 cup	lemon juice	125 mL
1 tbsp	dark sesame oil	15 mL
2	cloves garlic, minced	2
1 tsp	salt	5 mL
1/2 tsp	ground cumin	2 mL
1/2 tsp	ground coriander	2 mL
3/4 cup	toasted slivered almonds (for technique see Tip, page 54)	175 mL
1/2 cup	chopped walnuts (optional)	125 mL

1. In a large bowl, combine quinoa, wild rice, tomatoes, parsley, figs and green onions.

2. In a small bowl, whisk together lemon juice, sesame oil, garlic, salt, cumin and coriander. Pour over grains and vegetables; toss together. Refrigerate salad until cold. Sprinkle with almonds and walnuts just before serving.

To cook rice

Place 1/4 cup (50 mL) dry rice in 1 cup (250 mL) water. Bring to a boil. Reduce heat, cover and simmer for 60 minutes (water should be absorbed by the rice). Remove from heat and allow to cool.

This recipe is a delicious blend of grains and nuts that will enhance any meal. You'll find that this calcium-rich dish is a hearty and flavorful crunchy starter, thanks to the figs, almonds and walnuts.

CALCIUM QUOTE

"The African Bantu woman('s)... diet is free of milk and still provides 250 to 400 mg of calcium from plant sources...[they] commonly have 10 babies during their life and breast feed each of them.... But even with this huge calcium drain and relatively low calcium intake, osteoporosis is relatively unknown... ."

John A. McDougall, M.D. Author of *The McDougall Plan*

PER SERVING								
Calcium	Calories	Protein	Fat	Saturated Fat	Carbohydrates	Fiber	Iron	Sodium
121 mg	326	9 g	13 g	1.4 g	49 g	7 g	5 mg	407 mg

Quinoa Tabouli

SERVES 4 TO 6

2 1/2 cups	cooked quinoa (see page 20)	625 mL
3/4 cup	chopped mint	175 mL
1/2 cup	diced seedless cucumbers	125 mL
2 cups	finely chopped parsley	500 mL
2	small tomatoes, diced	2
3	green onions, chopped	3

Dressing

3 tbsp	lemon juice	45 mL
1 tbsp	olive oil	15 mL
1/2 tsp	salt	2 mL
1/4 tsp	black pepper	1 mL

1. In a bowl combine quinoa, mint, cucumbers, parsley, tomatoes and onions.

2. In a small bowl, whisk together the lemon juice, oil, salt and pepper. Pour over salad; toss. Serve at room temperature or refrigerate 1 hour and serve cold.

Tabouli is a popular Middle Eastern grain salad traditionally made with cracked wheat. Here we substitute quinoa to provide a similar flavor but a lot more calcium. We often serve this colorful, refreshing and light salad before our main course.

TIPS

This dish tastes best if it sits at least 5 minutes before serving to allow the quinoa to absorb the seasonings.

For an interesting change, try this salad as a stuffing for tomatoes.

				PER SERVING (4)				
Calcium	Calories	Protein	Fat	Saturated Fat	Carbohydrates	Fiber	Iron	Sodium
94 mg	238	8 g	7 g	0.8 g	39 g	5 g	7 mg	319 mg

Sesame Orange Dressing

MAKES 1 3/4 CUPS (425 ML)

4 oz	soft tofu	125 g
1/2 cup	orange juice (preferably calcium fortified)	125 mL
1/4 cup	chopped green onions	50 mL
1/4 cup	chopped cilantro	50 mL
3 tbsp	lemon juice	45 mL
3 tbsp	sesame seeds	45 mL
2 tbsp	balsamic vinegar	25 mL
1 tbsp	olive oil	15 mL
1/2 tbsp	soy sauce	5 mL
2 1/2 tsp	prepared mustard	12 mL
2 tsp	maple syrup	10 mL
1 tsp	dark sesame oil	5 mL
2	cloves garlic, minced	2
1/2 tsp	minced ginger root	2 mL

The combination of fresh cilantro and ginger makes this tangy dressing sweetly aromatic and flavorful, while the tofu provides its creamy consistency.

1. In a blender or food processor, combine tofu and orange juice; blend until creamy.

2. In a bowl, whisk together remaining ingredients. Add to tofu mixture; blend together a few seconds. Refrigerate until cool or use immediately.

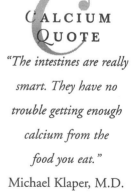

CALCIUM QUOTE

"The intestines are really smart. They have no trouble getting enough calcium from the food you eat."

Michael Klaper, M.D.

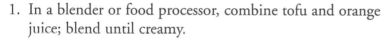

PER 2 TBSP (25 mL)

Calcium	Calories	Protein	Fat	Saturated Fat	Carbohydrates	Fiber	Iron	Sodium
31 mg	40	1 g	3 g	0.4 g	3 g	0.2 g	0.3 mg	52 mg

Maple Poppy Seed Dressing

MAKES 1 1/2 CUPS (375 mL)

Try this sweet tofu-based dressing on any salad. The poppy seeds are an interesting and tasty addition to this calcium-rich dressing.

6 oz	regular tofu or soft tofu	175 g
1/4 cup	balsamic vinegar	50 mL
3 tbsp	olive oil	45 mL
1/4 cup	maple syrup	50 mL
1 tbsp	lemon juice	15 mL
1 tsp	tahini	5 mL
1 tsp	granulated sugar	5 mL
1	clove garlic, minced	1
1/2 tsp	Dijon mustard	2 mL
2 tsp	poppy seeds	10 mL

1. In a blender or food processor, combine tofu, vinegar, oil, maple syrup, lemon juice, tahini, sugar, garlic and mustard; blend until creamy.

2. Pour dressing into a bottle or small container; add in poppy seeds and combine. Refrigerate 1 hour before serving.

WILD RICE QUINOA SALAD (PAGE 61) ➤
OVERLEAF: YELLOW AND GREEN BEAN SALAD WITH ARTICHOKES (PAGE 60)

PER 2 TBSP (25 mL)								
Calcium	Calories	Protein	Fat	Saturated Fat	Carbohydrates	Fiber	Iron	Sodium
28 mg	69	1 g	4 g	0.6 g	6 g	0.2 g	1 mg	5 mg

Main Courses

*I*n this section you'll discover tantalizing calcium-rich main courses suitable for a variety of meals — from light and simple to elegant and extravagant.

Many of these tasty recipes include tofu, a wonderful ingredient that absorbs the flavor of anything you cook with it. Even better, it provides a big calcium boost to the entire meal.

While these dishes are notable for their calcium content, your family and guests will find them comfortingly familiar. Main courses such as scrambled "eggs", quiche, perogies, chili, casseroles, and more, provide a delicious variety that everyone can enjoy.

< TOFU AND KALE QUICHE (PAGE 67)

Simple Scrambled "Eggs"

SERVES 1 OR 2

Kids and adults alike will appreciate this dish, since it looks and feels like eggs but is actually tofu in disguise. For an interesting variation, try adding chopped red pepper or other favorites, such as sliced mushrooms.

1 tsp	margarine	5 mL
1/4 cup	chopped green onions	50 mL
8 oz	firm tofu, crumbled	250 g
1/4 tsp	salt	1 mL
1/4 tsp	garlic salt	1 mL
1/8 tsp	turmeric	0.5 mL
1/8 tsp	ground cumin	0.5 mL
Pinch	black pepper	Pinch

1. In a small skillet, melt margarine over medium heat. Add green onions and sauté 3 to 4 minutes. Add remaining ingredients and scramble until desired "egg-like" consistency is reached, about 4 minutes. If too dry, add a few drops of water.

TIP

To boost the calcium, add finely chopped steamed broccoli or bok choy.

PER SERVING (1)

Calcium	Calories	Protein	Fat	Saturated Fat	Carbohydrates	Fiber	Iron	Sodium
247 mg	187	18 g	12 g	1.7 g	6 g	1 g	12 mg	462 mg

Tofu and Kale Quiche

SERVES 4 TO 6

PREHEAT OVEN TO 350° F (180° C)

1/2 tsp	canola oil	2 mL
1/2 cup	chopped green onions	125 mL
8 oz	firm tofu, crumbled	250 g
8 oz	soft tofu, crumbled	250 g
1 cup	packed finely chopped kale or turnip greens	250 mL
1 cup	grated soy cheese	250 mL
1/2 cup	chopped red bell peppers	125 mL
1/2 tsp	salt	2 mL
1/2 tsp	turmeric	2 mL
	9-inch (22.5 cm) store-bought pastry pie shell	

Here's a novel interpretation of traditional egg-based quiche that uses tofu to create a tasty, calcium-rich version of the original. The bonus in this recipe is the kale, which provides an additional calcium boost.

1. In a small nonstick skillet, heat oil over medium heat. Add green onions and sauté for 3 minutes. Add remaining ingredients to skillet and mix together until cheese begins to soften and kale wilts slightly.

2. Transfer mixture to prepared pie shell (see package instructions). Bake quiche in preheated oven for 40 minutes.

PER SERVING (4)

Calcium	Calories	Protein	Fat	Saturated Fat	Carbohydrates	Fiber	Iron	Sodium
381 mg	391	23 g	23 g	4.9 g	25 g	2 g	8 mg	611 mg

Greens and Grains Mushroom Casserole

SERVES 8 TO 10

PREHEAT OVEN TO 350° F (180° C)
13- BY 9-INCH (3 L) CASSEROLE DISH, LIGHTLY GREASED

1 cup	uncooked quinoa	250 mL
1 cup	uncooked brown rice	250 mL
6 cups	packed finely chopped collard greens	1.5 L
1 1/2 tsp	olive oil	7 mL
2	onions, chopped	2
2	stalks celery, finely chopped	2
2	carrots, finely chopped	2
2 cups	mushrooms, chopped or sliced	500 mL
3	cloves garlic, minced	3
1 tsp	garlic salt	5 mL
2 tsp	salt	10 mL
1 tsp	black pepper	5 mL
2 cups	soy milk (calcium fortified)	500 mL
1/3 cup	all-purpose flour	75 mL
2 1/2 tbsp	soy sauce	35 mL
1 lb	firm tofu, crumbled	500 g
1/2 cup	ground almonds	125 mL

1. Rinse quinoa and rice using a fine strainer under cold water. Place both in a medium-sized pot with 4 cups (1 L) water. Cover, bring to a boil, reduce heat and simmer 25 to 35 minutes or until water is absorbed. Transfer grains to a large bowl.

2. Steam collards until soft. Remove from heat and add to bowl of grains.

3. In a large nonstick skillet, heat oil over medium heat; add onions, celery, carrots, mushrooms, garlic, garlic salt, salt and pepper; cover and cook 10 minutes, stirring occasionally. Add soy milk and flour; cook, stirring constantly, until thick. Transfer to bowl of grains.

4. Add soy sauce, crumbled tofu and almonds to vegetable mixture. Stir until combined and press mixture into casserole dish. Bake, uncovered, for 1 hour.

PER SERVING (8)								
Calcium	Calories	Protein	Fat	Saturated Fat	Carbohydrates	Fiber	Iron	Sodium
354 mg	383	20 g	13 g	1.6 g	53 g	7 g	10 mg	1103 mg

Whether you're looking for a tasty main course or even a side dish, this casserole is the perfect choice. The soft consistency and light mushroom flavor is a welcome change. It's also loaded with calcium!

Squash with Sweet Veggie Stuffing

SERVES 4

PREHEAT OVEN TO 400° F (200° C)

2	*small butternut squash*	2
2 1/2 cups	*finely chopped broccoli*	625 mL
2	*parsnips, grated*	2
2	*carrots, grated*	2
2 cups	*packed finely chopped kale*	500 mL
1 1/2 tbsp	*olive oil*	20 mL
1/4 cup	*maple syrup*	50 mL
2 tbsp	*water*	25 mL
1/2 tsp	*cinnamon*	2 mL
1/4 tsp	*salt*	1 mL
1/8 tsp	*black pepper*	0.5 mL

1. Cut squash in half and remove seeds. Wrap in foil and bake 50 to 60 minutes. Remove from oven and set aside to cool.

2. Meanwhile, in a large bowl, combine broccoli, parsnips, carrots and kale. Toss together.

3. In a small bowl, whisk together the oil, maple syrup, water, cinnamon, salt and pepper. Pour over broccoli mixture; toss until well coated.

4. In a large nonstick skillet or wok over medium-high heat, sauté broccoli mixture for 5 minutes or until vegetables are tender-crisp. Remove from skillet and set aside.

5. Hollow out center of cooled squash, leaving a thickness of about 1/4 inch (5 mm). Fill with vegetable mixture.

6. Place stuffed squash in a baking dish filled with 3/4 inch (2 cm) water. (Use 2 baking dishes if necessary.) Cover with foil and bake in preheated oven for 30 minutes.

Here's a great way to get kids to eat calcium-rich veggies! All you have to do is sauté delicious, colorful vegetables in a maple syrup-based mixture then stuff the squash to create a uniquely sweet and satisfying dish.

CALCIUM QUOTE

"If you moderate your [animal] protein intake and limit your intake of sodium and salty foods, your calcium needs will be lower than those of the typical [person]."
Suzanne Havala M.S., R.D.
Dietician, researcher
and author

PER SERVING								
Calcium	Calories	Protein	Fat	Saturated Fat	Carbohydrates	Fiber	Iron	Sodium
254 mg	325	7 g	6 g	0.9 g	69 g	11 g	4 mg	218 mg

Chunky Chili

SERVES 8

This sensational, lightly spiced, chunky chili combines tofu, chickpeas and beans. If you love spicy dishes, simply increase the amount of chili powder. You can spoon this chili onto an open-faced bun and enjoy it as a "sloppy joe." As an added-calcium variation, we like to pour this chili over cooked quinoa.

1	can (5 1/2 oz [156 mL]) tomato paste	1
1/3 cup	dry red wine	75 mL
2 tbsp	Dijon mustard	25 mL
2 tbsp	soy sauce	25 mL
1 tbsp	dried basil	15 mL
2 1/2 tsp	dried oregano	12 mL
4	cloves garlic, chopped	4
1 1/2 lbs	firm tofu, crumbled	750 g
2	cans (each 28 oz [796 mL]) plum tomatoes, with juice, chopped	2
2 tbsp	chili powder	25 mL
1 tbsp	olive oil	15 mL
2	small onions, chopped	2
2 cups	cooked black beans or 1 can (19 oz [540 mL]) rinsed and drained	500 mL
1 1/2 cups	cooked chickpeas or 1 can (14 oz [398 mL]) rinsed and drained	375 mL
1/2 cup	chopped parsley	125 mL
2 tbsp	dried coriander	25 mL

1. In a bowl combine tomato paste, red wine, mustard, soy sauce, basil, oregano and garlic; mix well. Stir in crumbled tofu. Set aside.

2. In a large pot, combine tomatoes and chili powder. Simmer over low heat, covered, for 5 minutes.

3. Meanwhile, in a large nonstick skillet, heat oil over medium heat. Add onions and sauté 2 minutes. Add tofu mixture and cook for 1 more minute. Transfer to tomato-chili mixture in pot; simmer, stirring occasionally, for 40 minutes.

4. Add beans, chickpeas, parsley and coriander. Simmer 10 minutes.

PER SERVING

Calcium	Calories	Protein	Fat	Saturated Fat	Carbohydrates	Fiber	Iron	Sodium
311 mg	334	24 g	11 g	1.6 g	41 g	9 g	14 mg	689 mg

Hearty Bean Gumbo

SERVES 6

1 tbsp	olive oil	15 mL
1	large onion, finely chopped	1
4	cloves garlic, minced	4
2 cups	cooked soybeans or adzuki beans or *1 can (19 oz [540 mL])* rinsed and drained	500 mL
3 cups	okra, sliced into rounds 1/4 inch (5 mm) thick	750 mL
2 cups	diced tomatoes	500 mL
3/4 cup	vegetable stock	175 mL
3/4 cup	ground almonds	175 mL
1/2 cup	finely chopped red bell peppers	125 mL
1/4 cup	finely chopped parsley	50 mL
1/4 cup	lemon juice	50 mL
1	bay leaf	1
1 tsp	dried basil	5 mL
1/2 tsp	salt	2 mL
1/4 tsp	black pepper	1 mL
1/8 tsp	hot pepper sauce (optional)	0.5 mL

1. In a large nonstick skillet with a lid, heat oil over medium heat. Add onion and garlic; sauté 5 minutes.

2. Add remaining ingredients to skillet. Reduce heat to low and simmer, covered, for 35 minutes, stirring occasionally. Remove bay leaf before serving.

Try serving this flavorful gumbo stuffed into red or yellow bell peppers or seeded tomatoes to create a hearty calcium-rich meal. You can also serve this dish on its own or, to boost the calcium, mix together with cooked grains such as quinoa.

TIP

To stuff peppers or tomatoes: cut off top 1/4 inch (5 mm); remove and discard seeds. Fill shells with cooked mixture and set in a deep casserole dish. Add 1/4 inch (5 mm) water and bake, covered, for 30 minutes in a 350° F (180° C) oven.

				PER SERVING				
Calcium	Calories	Protein	Fat	Saturated Fat	Carbohydrates	Fiber	Iron	Sodium
146 mg	241	14 g	14 g	1.7 g	19 g	7 g	4 mg	279 mg

Tasty Tofu Tacos

MAKES 15 TACOS

Here's an unusual and flavorful way to serve calcium-rich tofu. Use hard taco shells or try rolling the filling in a soft tortilla. Either way, your family will enjoy this creative meal.

1 tbsp	vegetable oil	15 mL
2	onions, finely chopped	2
2	cloves garlic, minced	2
1 lb	extra firm tofu, crumbled	500 g
2 tbsp	soy sauce	25 mL
1 cup	mushroom pasta sauce	250 mL
1 tbsp	chili powder	15 mL
2 tsp	dried basil	10 mL
1/2 tsp	salt	2 mL
3 cups	fresh or frozen okra, sliced	750 mL
15	taco shells	15

1. In a large nonstick skillet with a lid, heat oil over medium heat. Add onions and garlic; sauté for 3 minutes. Add tofu and soy sauce; cook, stirring occasionally, another 8 minutes. Add pasta sauce, chili powder, basil and salt; cook 3 minutes.

2. Reduce heat to low. Add okra and simmer, covered, for 10 minutes or until okra is soft. Stir occasionally.

3. Scoop into individual taco shells and serve hot.

CALCIUM BOOSTER

Try taco shells made with lime, as they contain added calcium. You can also add shredded, lightly steamed kale to boost the calcium or simply add shredded lettuce. Top with grated soy cheese for a richer meal.

			PER SERVING					
Calcium	Calories	Protein	Fat	Saturated Fat	Carbohydrates	Fiber	Iron	Sodium
118 mg	134	7 g	5 g	0.2 g	16 g	3 g	1 mg	305 mg

Colorful Vegetable Bake

SERVES 8

PREHEAT OVEN TO 350° F (180° C)
13- BY 9-INCH (3.5 L) BAKING DISH, LIGHTLY GREASED

5	medium potatoes, peeled and quartered	5
2 cups	chopped tomatoes	500 mL
1 1/4 cups	ground almonds	300 mL
1 cup	finely chopped Brussels sprouts	250 mL
1 cup	packed chopped dandelion or collard greens	250 mL
1 cup	finely chopped onions	250 mL
2 cups	finely chopped broccoli	500 mL
8 oz	firm tofu, crumbled	250 g
1/4 cup	chopped dill	50 mL
3	stalks celery, finely chopped	3
3	carrots, finely chopped	3
2 tbsp	soy sauce	25 mL
1 tbsp	olive oil	15 mL
3	cloves garlic, minced	3
1 tbsp	dried basil	15 mL
1 1/4 tsp	salt	6 mL

1. Boil potatoes until soft. Drain and mash.

2. In a large bowl, combine remaining ingredients with mashed potatoes; mix well. Firmly pack mixture into prepared dish and bake, covered, in preheated oven for 1 1/2 hours. Uncover and bake for another 10 minutes. Serve hot.

This lightly seasoned colorful casserole is excellent as a main course or side dish. Although a little time-consuming to prepare, this recipe has plenty of calcium and is well worth the effort. Try serving it with a warm calcium-rich tomato sauce drizzled over top (see Pasta section, page 115, for sauce ideas).

CALCIUM FACT

According to some research, preventing calcium loss is 3 to 4 times more important than calcium intake.

PER SERVING

Calcium	Calories	Protein	Fat	Saturated Fat	Carbohydrates	Fiber	Iron	Sodium
172 mg	273	12 g	13 g	1.5 g	32 g	6 g	5 mg	680 mg

Shepherd's Pie

SERVES 6 TO 8

PREHEAT OVEN TO 350° F (180° C)
13- BY 9-INCH (3.5 L) CASSEROLE DISH, LIGHTLY GREASED

This shepherd's pie not only looks impressive, but is a tasty, calcium-rich alternative to an old favorite. It's also the one our family enjoys the most on cold winter nights.

1 1/2 cups	cooked soybeans or navy beans or *1 can (14 oz [398 mL])* rinsed and drained	375 mL
1 cup	pasta sauce	250 mL
1/4 cup	soy sauce	50 mL
2 tbsp	olive oil	25 mL
3 cups	coarsely chopped broccoli florets	750 mL
1 1/2 cups	corn niblets	375 mL
2 cups	packed chopped mustard greens or *Chinese lettuce*	500 mL
1 cup	sliced okra	250 mL
1	medium onion, chopped	1
4	carrots, finely chopped	4
1	potato, peeled and chopped	1
2 tsp	granulated sugar	10 mL
2 tsp	dried basil	10 mL

Potato topping

5	medium potatoes, peeled	5
1 tbsp	olive oil	15 mL
1/4 cup	soy milk (calcium fortified)	50 mL
1/2 tsp	salt	2 mL
	Paprika	

1. In a large bowl, using a potato masher (or in a food processor), roughly mash beans. Stir in pasta sauce and soy sauce. Set aside.

PER SERVING (6)								
Calcium	Calories	Protein	Fat	Saturated Fat	Carbohydrates	Fiber	Iron	Sodium
165 mg	363	15 g	11 g	1.6 g	56 g	10 g	5 mg	1188 mg

2. In a large nonstick skillet, heat oil over medium heat. Add remaining ingredients and cook, stirring occasionally, for 8 to 10 minutes. Add mashed bean mixture and stir until well combined.

3. Make the potato topping: In a pot of boiling water, cook potatoes until soft. Place in bowl and mash together with olive oil, soy milk and salt.

4. Transfer bean mixture to prepared casserole dish and spread out evenly. Add potato topping and smooth out. Sprinkle with paprika and bake 40 minutes.

CALCIUM FACT

Like any disease, it's far
better to prevent
osteoporosis than
to try to treat it
later in life.

Spanakopita

MAKES 18 TO 20 TRIANGLES

PREHEAT OVEN TO 375° F (190° C)

Spanakopita is a popular Greek pie traditionally made from cheese and spinach. The filling is wrapped in phyllo pastry and can be formed into any size and shaped into triangles, rectangles or balls. This recipe is a great reproduction of the popular, traditional spanakopita, only the phyllo pastry here is wrapped around a delicious blend of high-calcium greens, tofu and soy cheese.

2 cups	finely chopped kale	500 mL
1 cup	finely chopped collard greens	250 mL
1 tbsp	olive oil	15 mL
1 cup	finely chopped onions	250 mL
8 oz	firm tofu, crumbled	250 g
4	cloves garlic, minced	4
1 1/4 cups	shredded soy cheese	300 mL
1/2 cup	finely chopped dill	125 mL
1/2 cup	finely chopped parsley	125 mL
1 tbsp	lemon juice	15 mL
3/4 tsp	salt	4 mL
1/4 tsp	nutmeg	1 mL
1/8 tsp	black pepper	0.5 mL
1	package phyllo dough	1

1. Steam kale and collards until soft. Let cool and squeeze out excess liquid. Place in a large bowl.

2. In a large nonstick skillet, heat oil over medium heat. Add onions, tofu and garlic; sauté for 5 minutes. Add to bowl of greens, along with soy cheese, dill, parsley, lemon juice, salt, nutmeg and black pepper; mix well. Set aside.

3. Place 2 to 4 tbsp (25 mL to 50 mL) olive oil in a small dish. Set beside your work area. Use this oil to brush phyllo sheets.

				PER TRIANGLE				
Calcium	Calories	Protein	Fat	Saturated Fat	Carbohydrates	Fiber	Iron	Sodium
79 mg	112	6 g	6 g	0.7 g	10 g	1 g	2 mg	193 mg

4. Lay one phyllo sheet flat on wax paper. Lightly brush some oil across the sheet of phyllo then lay another phyllo sheet on top of the first one. (Remember that phyllo dough dries out quickly so keep the phyllo sheets in the plastic wrap until you are ready to use each sheet.)

5. With a sharp knife or clean scissors, make 4 even strips. (Each strip will become a complete triangle.) Place 1 heaping tbsp (20 mL) mixture on bottom right corner of strip. Fold bottom right corner over mixture toward the left to form a triangle. Brush a little oil over top of triangle. Continue to fold into a triangle shape (brushing a little oil over top and edges) until the strip of phyllo is used up.

6. Place on greased baking sheet and lightly brush tops of triangles with oil. Bake until golden brown and crisp, approximately 15 to 20 minutes.

*T*IP

Fun with phyllo
You can experiment
with your own elegant
stuffed phyllo designs
by making rolls, balls
with fancy tops or other
interesting shapes.

Perogies

MAKES 20 TO 25 PEROGIES

PREHEAT OVEN TO 350°F (180°C)

Perogies are a tasty Ukrainian dish that has recently become popular in North America. Traditionally, these soft dough-filled pockets are stuffed with potatoes or cheese. But in this unique recipe, we stuff the pockets with high-calcium greens and Chinese lettuce. We add mock bacon bits to give the perogies a pleasing smoked flavor and aroma. They're fantastic to serve at a dinner party to complement other dishes, or as a main course with a hearty soup or salad.

Filling

1	medium sweet potato	1
2 cups	finely chopped Chinese lettuce	500 mL
1 cup	finely chopped beet greens, turnip greens or kale (ribs removed)	250 mL
1 tbsp	olive oil	15 mL
2	large onions, finely chopped	2
1 cup	cooked quinoa (page 20)	250 mL
1 1/2 tbsp	simulated bacon bits	20 mL
2	cloves garlic, minced	2
1/2 tsp	minced ginger root	2 mL
1/2 tsp	salt	2 mL
1/4 tsp	chili flakes	1 mL
1/3 cup	ground almonds	75 mL

Dough wrap

2 1/4 cups	all-purpose flour	550 mL
1 tsp	salt	5 mL
1 tbsp	margarine, melted	15 mL
3/4 cup	water	175 mL
1 tsp	cider vinegar	5 mL

1. Pierce potato with fork, wrap in foil and bake in preheated oven for 60 to 70 minutes or until soft. Peel off skin and mash potato in a small bowl. Set aside.

2. Steam lettuce and beet greens until tender. Let cool; squeeze out excess liquid and set aside.

PER PEROGIE

Calcium	Calories	Protein	Fat	Saturated Fat	Carbohydrates	Fiber	Iron	Sodium
25 mg	107	3 g	2 g	0.3 g	19 g	2 g	1 mg	212 mg

3. In a large nonstick skillet, heat oil over medium heat. Add onions and sauté until soft. Add mashed sweet potato, lettuce, greens, quinoa, bacon bits, garlic, ginger, salt and chili flakes. Cook, stirring occasionally, for 5 minutes. Stir in almonds; remove from heat. Transfer mixture to a bowl and allow to cool.

4. Make the dough wrap: In a large bowl, combine flour and salt. Drizzle with melted margarine. Add water and vinegar; mix together to create a slightly firm dough. Knead a little by hand, adding a little flour if the dough is too sticky.

5. Dust a large work surface and rolling pin with flour. Roll out dough to about 1/8 inch (2.5 mm) thick. Using a drinking glass or round cookie cutter, cut 3-inch (7.5 cm) circles of dough.

6. Place 1 tbsp (15 mL) of filling onto each circle and stretch one side over to meet the other, forming a half circle. If necessary, moisten the edges with water and press together to seal. (Make sure the perogies are completely sealed or the mixture will leak out when you boil them.) For a decorative effect, lightly press the edge of the dough around the perogie with the handle of a flat utensil to make fancy wedges in the dough. (Don't press too hard though, or you will cut into the perogie.)

7. To a large pot of boiling water, carefully add perogies in batches of 8 to 10. Cook 2 to 3 minutes or until perogies float to the top. (They should have a rubbery texture.)

8. Dab cooked perogies with margarine and set aside. In a nonstick skillet, fry perogies 2 minutes per side until golden brown. Serve hot perogies plain or with your favorite topping.

TIPS

During preparation, keep excess dough in a plastic bag to keep it from drying out. Also, as perogies are assembled, but before they are cooked, cover with a slightly damp towel to keep them moist.

Perogies can also be baked instead of fried. Place them on a lightly oiled baking sheet. Brush each with a little oil and bake 10 minutes on one side and 5 minutes on the other in a preheated 350° F (180° C) oven.

Stuffed Collard Greens

MAKES 20 TO 25 ROLLS

PREHEAT OVEN TO 350° F (180° C)

If you enjoy Middle Eastern stuffed grape leaves, then you'll love this variation, which uses collard leaves wrapped around quinoa and rice. It's a delicious calcium-rich alternative to traditional vine leaves stuffed with rice.

20 to 25	collard green leaves, rinsed, thick stems trimmed	20 to 25
1/3 cup	olive oil	75 mL
1/4 cup	balsamic vinegar	50 mL
2 tbsp	lemon juice	25 mL
2 tsp	dried thyme	10 mL
4	cloves garlic, minced	4
1/2 tsp	salt	2 mL

Stuffing

1/4 cup	olive oil	50 mL
2	onions, finely chopped	2
4	cloves garlic, minced	4
3 cups	water	750 mL
1 cup	uncooked quinoa, rinsed	250 mL
1/2 cup	uncooked white rice	125 mL
3	lemons, juice only	3
1/4 cup	finely chopped dill	50 mL
3/4 cup	finely chopped mint	175 mL
1 tsp	paprika	5 mL
1 tsp	salt	5 mL
1/2 tsp	allspice	2 mL
1/2 tsp	black pepper	2 mL
1	cinnamon stick	1

TIP

Use any leftover marinade for drizzling over the rolled leaves when serving.

1. In large pot with a lid, steam collard leaves, 5 to 10 at a time, in 1/4 inch (5 mm) water for 1 minute or until slightly wilted. Drain and set aside.

				PER ROLL (20)				
Calcium	Calories	Protein	Fat	Saturated Fat	Carbohydrates	Fiber	Iron	Sodium
63 mg	101	2 g	4 g	0.6 g	14 g	2 g	1 mg	140 mg

2. Make the marinade: In a large saucepan over medium-low heat, combine oil, vinegar, lemon juice, thyme, garlic and salt; cook 5 minutes and remove from heat.

3. Dip collard leaves into marinade and transfer to a large bowl. Pour any leftover marinade over leaves. Set aside.

4. Make the stuffing: In a large pot with a lid, heat oil over medium heat. Add onions and sauté for 4 minutes. Add garlic and sauté 1 minute. Add remaining ingredients, including uncooked quinoa and rice. Bring to a light boil, reduce heat and simmer, covered, 20 to 25 minutes or until all liquid is absorbed. Discard cinnamon stick and remove from heat.

5. Place 2 to 3 tbsp (25 to 45 mL) of filling in center of each collard leaf. (Use more or less filling, depending on size of each leaf.) Fold bottom of leaf over filling, then fold sides in towards center, then roll top of leaf over to encase filling.

6. Place each roll on lightly greased baking sheet, seam side down. Bake 15 to 20 minutes.

Four-Season Savory Stew

SERVES 6 TO 8

This well seasoned, hearty stew is one of our favorites because it is simple to prepare and can be made hours in advance and kept warm throughout the day as it cooks slowly in a crockpot. (This is a slow cooking pot that can be left on overnight so that the beans and other ingredients cook completely on their own.) Serve this delicious stew by itself or spoon over cooked rice, quinoa or pasta shells — it's so versatile!

1 cup	dried black or white beans or soybeans or navy beans	250 mL
1 cup	dried chickpeas	250 mL
1 cup	dried quinoa	250 mL
2 cups	chopped rapini	500 mL
6 cups	water	1.5 L
1	can (28 oz [796 mL]) whole plum tomatoes	1
4	plum tomatoes, cut into quarters	4
4	medium potatoes, peeled and chopped into large chunks	4
1	sweet potato, peeled and chopped into chunks	1
1 cup	coarsely chopped Spanish onion	250 mL
3	bay leaves	3
2 tbsp	canola oil	25 mL
2	garlic cloves, peeled	2
2 1/2 tsp	salt	12 mL
2 tsp	turmeric	10 mL
1 tsp	ground coriander	5 mL
1 tsp	paprika	5 mL
1/2 tsp	cayenne pepper	2 mL

TIP

If you don't have a crockpot, use a regular large pot with a lid. Place in preheated 250° F (120° C) oven for 11 hours. Stir twice during cooking time. Add water if necessary.

1. If preparing stew for dinner the same day: Start early, since you'll need about 11 hours of cooking time. Add all ingredients to a large crockpot and stir gently until well mixed. Turn crockpot on high and cook for 3 1/2 hours. Reduce heat to low; cook, stirring occasionally, for remaining 7 1/2 hours or until beans are soft. Serve hot.

PER SERVING (6)								
Calcium	Calories	Protein	Fat	Saturated Fat	Carbohydrates	Fiber	Iron	Sodium
256 mg	550	22 g	9 g	0.9 g	100 g	15 g	10 mg	1208 mg

2. *If preparing stew the day before for the following day's meal:*
   Add all ingredients to a large crockpot in the late afternoon
   or evening, gently mix together. Turn crockpot on high and
   cook for 2 1/2 hours. Reduce heat to low and cook overnight.
   If serving the stew over rice or other grains, prepare ahead of
   serving time. Spoon hot stew over grains and serve.

Lemon Ginger Stir-Fry with Almonds

SERVES 6

Here's a sweet and simple way to add calcium-rich vegetables to your diet. This dish gets an added calcium boost when maple syrup is added and topped with almonds. You can use the vegetables specified here or select other high-calcium alternatives. (See Calcium Table, page 174.)

1 tbsp	dark sesame oil	15 mL
1 cup	halved yellow string beans	250 mL
1 cup	chopped rapini	250 mL
2 cups	chopped broccoli	500 mL
2	leeks, coarsely chopped	2
4	cloves garlic, minced	4
1 tbsp	cornstarch	15 mL
1 cup	water	250 mL
1 tbsp	lemon juice	15 mL
1 tbsp	grated ginger root	15 mL
3 tbsp	tamari sauce	45 mL
1 tbsp	soy sauce	15 mL
2 tsp	granulated sugar	10 mL
4 cups	finely chopped Chinese lettuce	1 L
2 cups	roughly chopped bok choy	500 mL
2	red bell peppers, chopped	2
5 cups	cooked quinoa (see page 20)	1.25 L
3/4 cup	slivered almonds, toasted (see Tip, page 54, for technique)	175 mL
1/4 cup	maple syrup	50 mL

TIP

Rice can replace quinoa. (See tip, facing page.)

1. In a large nonstick skillet, heat oil over medium-high heat. Add string beans, rapini, broccoli, leeks and garlic; stir-fry for 7 minutes. (Add 2 tbsp [25 mL] water if stir-fry becomes a little dry.) Set aside.

2. In a frying pan or wok over medium heat, combine cornstarch and water; cook, stirring, until it thickens slightly. Add lemon juice, ginger, tamari, soy sauce and sugar. Cook 3 minutes. Raise heat to medium-high. Add cooked string beans/rapini/broccoli mixture, Chinese lettuce, bok choy and red peppers; cook 8 minutes, stirring often.

3. Toss with quinoa and maple syrup. Garnish with a sprinkling of toasted almonds.

				PER SERVING				
Calcium	Calories	Protein	Fat	Saturated Fat	Carbohydrates	Fiber	Iron	Sodium
217 mg	455	16 g	15 g	1.6 g	70 g	9 g	9 mg	725 mg

Sesame Bok Choy and Carrot Stir-Fry

SERVES 2

1 tsp	dark sesame oil	5 mL
4	cloves garlic, minced	4
3	carrots, cut diagonally into 1/4-inch (5 mm) slices	3
1/2 cup	chopped green onions	125 mL
5 cups	bok choy, cut into 1/2-inch (1 cm) pieces	1.25 L
1/4 cup	vegetable stock	50 mL
2 tsp	minced ginger root	10 mL
1 tsp	granulated sugar	5 mL
2 tbsp	toasted sesame seeds	25 mL
3 cups	cooked quinoa (see page 20)	750 mL

1. In a large nonstick skillet or wok, heat oil over medium heat. Add garlic, carrots and green onions; stir-fry for 3 minutes. Add bok choy and stir-fry another 2 minutes. Stir in vegetable stock, ginger and sugar. Reduce heat and simmer 5 minutes.

2. Sprinkle sesame seeds over stir-fry. Spoon over quinoa.

Try this crunchy and colorful dish over a calcium-rich grain such as quinoa. This dish is lightly gingered with a hint of sesame.

TIP

Rice can replace quinoa as an accompaniment, although you'll get less calcium. Use 2 cups (500 mL) brown rice, rinsed and drained; add to 3 cups (750 mL) water in a large pot. Bring to boil; reduce heat, cover and simmer for 45 minutes.

PER SERVING

Calcium	Calories	Protein	Fat	Saturated Fat	Carbohydrates	Fiber	Iron	Sodium
324 mg	602	22 g	15 g	1.8 g	101 g	14 g	14 mg	260 mg

Side Dishes

Many of the side dishes here will be familiar. Others will be new to you. Old favorites such as mashed potatoes and tomato sauces may include beans, tofu or greens to boost the calcium. Our other unique recipes are based on calcium-rich ingredients such as quinoa, collards or kale.

Traditionally, side dishes were served to complement a main course. Nowadays, with so many people changing their eating habits to low-fat or vegetarian, "side dishes" are often the whole meal.

Our family relies on many of these tasty, calcium-rich recipes because they are so versatile. Our main meals often consist of a couple of hearty side dishes, which satisfies our individual tastes and always makes the meal more interesting.

Stuffed Tomatoes with Crumbled Rapini and Broccoli

SERVES 4

PREHEAT OVEN TO 350° F (180° C)
SHALLOW BAKING DISH

When added to rapini and broccoli, bread crumbs and chopped brazil nuts create a satisfying, crumbly texture — and helps to soften the strong flavor of calcium-rich rapini.

2 cups	chopped rapini (small pieces)	500 mL
3 1/2 cups	chopped broccoli (small pieces)	750 mL
3 tbsp	lemon juice	45 mL
1 tbsp	olive oil	15 mL
2 tsp	soy sauce	10 mL
1/4 tsp	curry powder	1 mL
1/3 cup	bread crumbs	75 mL
2 tbsp	finely chopped brazil nuts	25 mL
4	large tomatoes	4

1. In a large pot, steam or boil rapini and broccoli until tender-crisp. Drain and set aside.

2. In a large bowl, whisk together lemon juice, oil, soy sauce and curry powder. Add rapini and broccoli; toss to coat. Add bread crumbs and brazil nuts. Toss until combined. Set aside.

3. Slice 1/4-inch (5 mm) tops off tomatoes; remove flesh and seeds. Stuff tomatoes with the rapini mixture and place in a baking dish. Cover with foil and bake in preheated oven for 20 minutes or until soft.

PER SERVING

Calcium	Calories	Protein	Fat	Saturated Fat	Carbohydrates	Fiber	Iron	Sodium
136 mg	159	7 g	7 g	1.4 g	20 g	5 g	2 mg	282 mg

Moroccan-Style Carrots and Parsnips

SERVES 4

1 tsp	canola oil	5 mL
3 tbsp	brown sugar	45 mL
1 1/4 tsp	cinnamon	6 mL
1/8 tsp	ground cumin	0.5 mL
1 3/4 cups	orange juice (preferably calcium fortified)	425 mL
6	carrots, peeled and thinly sliced	6
8	parsnips, peeled and thinly sliced	8
1/2 cup	chopped dried figs	125 mL
1/4 cup	raisins	50 mL
Pinch	cayenne pepper	Pinch

1. In a large nonstick saucepan, heat oil over medium heat. Add brown sugar, cinnamon and cumin; cook, stirring, for 1 minute. Add orange juice, carrots, parsnips, figs, raisins and cayenne. Simmer, covered, for 20 to 25 minutes.

Sweet and cinnamony, this side dish goes well with almost any main course. Try over rice or quinoa, along with some of the cooking liquid, to turn this dish into a complete meal.

	PER SERVING							
Calcium	Calories	Protein	Fat	Saturated Fat	Carbohydrates	Fiber	Iron	Sodium
321 mg	456	7 g	3 g	0.4 g	109 g	15 g	4 mg	109 mg

Chinese Lettuce with Slivered Almonds and Caraway

SERVES 2 TO 4

If you love the flavor and crunch of toasted almonds, then here's a recipe for you. This versatile dish can be served either hot or cold. You can also try tossing it with a large green salad to add a boost of flavor.

1 tsp	dark sesame oil	5 mL
1 tsp	caraway seeds	5 mL
8 cups	chopped Chinese lettuce	2 L
1 1/2 tsp	dried thyme	7 mL
1 tsp	dried tarragon	5 mL
1 1/2 tsp	soy sauce	7 mL
1/2 tsp	salt	2 mL
1 tbsp	lemon juice	15 mL
1/4 cup	toasted slivered almonds (see Tip, page 54, for technique)	50 mL

1. In a large nonstick skillet or wok, heat oil over medium-low heat. Add caraway seeds and sauté for 1 minute. Add Chinese lettuce, thyme, tarragon, soy sauce and salt; stir-fry over medium-high heat for 5 to 6 minutes.

2. Add lemon juice and heat through for 1 minute. Sprinkle with toasted almonds and toss together; serve.

TIP

For added color, toss in 1/2 cup (125 mL) finely chopped, steamed red bell pepper.

		PER SERVING (2)						
Calcium	Calories	Protein	Fat	Saturated Fat	Carbohydrates	Fiber	Iron	Sodium
298 mg	171	7 g	11 g	1.3 mg	15 g	4 g	3 mg	864 mg

Bean and Kale Scramble

SERVES 4 TO 6

1 tsp	olive oil	5 mL
1	large onion, finely chopped	1
2 cups	Great Northern beans or white beans or *1 can (19 oz [550 g])* rinsed and drained	500 mL
1/4 cup	chopped dill	50 mL
5 cups	chopped kale	1.25 L
1/3 cup	freshly squeezed lemon juice	75 mL
3 tbsp	water	45 mL
2 tbsp	soy sauce	25 mL
1 tbsp	prepared mustard	15 mL

1. In a large nonstick skillet, heat oil over medium heat. Add onion and sauté for 3 minutes. Add cooked beans and dill; sauté another 3 minutes.

2. Add remaining ingredients and sauté 5 minutes. Stir occasionally. Serve hot or cooled to room temperature.

Here's an interesting side dish that mixes dark greens and beans with a savory lemon and dill flavor. It's a creative way to serve healthy calcium-rich greens.

	PER SERVING (4)							
Calcium	Calories	Protein	Fat	Saturated Fat	Carbohydrates	Fiber	Iron	Sodium
187 mg	183	11 g	2 g	0.4 g	33 g	6 g	4 mg	600 mg

Greens with Sun-Dried Tomatoes

SERVES 2

These high-calcium shredded greens are delicious served as a hot side dish. They're also terrific stuffed into seeded tomatoes (or small red or yellow bell peppers), then baked until soft. You'll find the end result is worth the extra effort.

1 tsp	olive oil	5 mL
3	cloves garlic, minced	3
1/4 cup	finely chopped onions	50 mL
3/4 cup	vegetable stock	175 mL
3 cups	finely chopped kale, ribs removed	750 mL
3 cups	finely chopped collard greens, ribs removed	750 mL
1/2 cup	roughly chopped drained sun-dried tomatoes (oil-packed variety)	125 mL

1. In a large nonstick skillet, heat oil over medium-high heat. Add garlic and onions; sauté for 2 minutes. Stir in vegetable stock and greens. Reduce heat to low and cook, stirring occasionally, for 8 minutes.

2. Add sun-dried tomatoes; toss with greens. Cook 2 minutes or until heated through. Serve hot.

TIP

Try cooling and tossing into a green salad to perk up the flavor while increasing the salad's calcium content.

				PER SERVING				
Calcium	Calories	Protein	Fat	Saturated Fat	Carbohydrates	Fiber	Iron	Sodium
237 mg	166	6 g	7 g	0.9 g	24 g	6 g	2 mg	351 mg

Gingered Broccoli

SERVES 4

2	bunches broccoli (about 2 lbs [1 kg]), chopped	2
1/2 tbsp	olive oil	5 mL
3 tbsp	orange juice (preferably calcium fortified)	45 mL
1 1/2 tbsp	soy sauce	20 mL
2 1/4 tsp	Dijon mustard	11 mL
1 tsp	brown sugar	5 mL
1/2 tsp	ground ginger	2 mL

This simple-to-make side dish is light and tasty. As the broccoli sautées, it softens and breaks up into smaller, more flavorful pieces.

1. Lightly steam broccoli until tender-crisp. Set aside.

2. In a saucepan, heat oil over medium heat. Add orange juice, soy sauce, mustard, brown sugar and ginger; simmer for 1 minute. Add broccoli and toss to coat. Simmer for 3 minutes. Serve hot.

TIP

To enhance the color of this dish, add 1/2 cup (125 mL) each of chopped steamed red and yellow bell peppers.

				PER SERVING				
Calcium	Calories	Protein	Fat	Saturated Fat	Carbohydrates	Fiber	Iron	Sodium
115 mg	79	7 g	2 g	0.3 g	13 g	6 g	2 mg	272 mg

Sesame Vegetable Bowl with Dill

SERVES 4 TO 6
PREHEAT OVEN TO 350° F (180° C)
13- BY 9-INCH (3.5 L) BAKING DISH

Here's a simple yet exciting way to enjoy a wide variety of calcium vegetables. A broad range of colors and flavors makes this dish even more appealing. Use the veggies in this recipe or add your own calcium-rich favorites. (Check the Calcium Table on page 174 for ideas.)

Marinade

1/2 cup	orange juice (preferably calcium fortified)	125 mL
2 tbsp	olive oil	25 mL
3 tbsp	sesame seeds	45 mL
4	cloves garlic, minced	4
3 tbsp	soy sauce	45 mL
1/4 cup	chopped dill	50 mL
1/2 tsp	black pepper	2 mL

Vegetables

Half	small rutabaga	Half
2	carrots	2
1	medium sweet potato	1
15	okra, trimmed and roughly chopped	15
4 cups	bok choy, chopped into 1-inch (2.5 cm) pieces	1 L
2 cups	broccoli florets, cut into bite-size pieces	500 mL

TIP

Some vegetables are harder than others and, as a result, take longer to cook. Try chopping and steaming the firmer vegetables first before baking.

1. Make the marinade: In a bowl, whisk together the orange juice, oil, sesame seeds, garlic, soy sauce, dill and pepper. Set aside.

2. Peel and chop rutabaga into 1/4-inch (5 mm) chunks (approximately 3 cups [750 mL] cubed). Steam 15 minutes. Set aside.

3. Peel carrots and sweet potato and chop into 1/4-inch (5 mm) pieces. Steam 10 minutes. Set aside.

4. Transfer steamed vegetables to marinade along with okra, bok choy and broccoli; toss to coat. Transfer to baking dish and cover with foil. Bake 30 minutes. Toss vegetables in marinade. Bake another 20 minutes. Serve hot.

PER SERVING (4)

Calcium	Calories	Protein	Fat	Saturated Fat	Carbohydrates	Fiber	Iron	Sodium
219 mg	216	7 g	6 g	0.8 g	37 g	8 g	3 mg	466 mg

Marinated Tofu Cubes over Quinoa

SERVES 2

PREHEAT OVEN TO 350° F (180° C)
11- BY 9-INCH (2.5 L) BAKING DISH

1/2 cup	any store-bought marinade (try barbecue sauce, hickory-smoked barbecue sauce, sweet-and-sour sauce, soy sauce, curry sauce or any of your favorite sauces)	125 mL
8 oz	firm tofu or extra firm tofu, cut into 1/4-inch (5 mm) cubes	250 g
1 cup	quinoa, cooked (see page 20 for technique)	250 mL

1. Place marinade in a bowl. Add tofu cubes and toss to coat. Cover and refrigerate for 2 hours to allow tofu to absorb flavors.

2. Transfer tofu to baking dish. Bake in preheated oven for 12 to 15 minutes.

3. Pour quinoa on top of cooked tofu and mix together. Serve hot.

These versatile and easy-to-make tofu cubes are great in a salad, added to a stir-fry or pasta dish, or placed on a skewer with calcium-rich vegetables. (After marinating, place cubes on a skewer with vegetables and bake in oven until vegetables turn soft.)

TIPS

For an interesting variation, try cutting firm tofu into square, flat slices, then marinate. These slices can be used as a filler in a sandwich.

Try simmering tofu cubes in marinade for 10 minutes over low heat in a nonstick skillet, covered.

PER SERVING

Calcium	Calories	Protein	Fat	Saturated Fat	Carbohydrates	Fiber	Iron	Sodium
268 mg	353	24 g	13 g	1.8 g	39 g	7 g	16 mg	534 mg

Tofu Baked Beans with Rich Savory Sauce

SERVES 6

Enjoy this hearty bean dish on its own, as a topping over pasta, or served "sloppy joe" style over a bun. The molasses not only boosts the calcium but also gives this dish its rich and delicious semi-sweet flavor.

1/2 tbsp	olive oil	5 mL
2	medium onions, chopped	2
1 cup	cooked navy beans or 1 can (14 oz [398 mL]) rinsed and drained	250 mL
1 1/2 cups	cooked black beans or 1 can (14 oz [398 mL]) rinsed and drained	375 mL
3	medium apples, peeled, cored and finely chopped	3
8 oz	firm tofu, cut into small cubes	250 g
1/4 cup	ketchup	50 mL
1/4 cup	blackstrap molasses	50 mL
1/4 cup	tomato paste	50 mL
1 tbsp	dry mustard	15 mL
1/2 tsp	cumin	2 mL

TIP

Try this dish with a baked potato! Pierce with fork, wrap in foil then bake potato in a preheated 425° F (220° C) oven for 50 to 60 minutes or until soft. Slice off top then hollow out with a spoon, leaving a thickness of 1/4 inch (5 mm). Stuff with baked beans and serve hot.

1. In a large nonstick saucepan, heat oil over medium heat. Add onions and sauté for 5 minutes or until softened. Add navy beans, black beans, apples, tofu, ketchup, molasses, tomato paste, mustard and cumin. Reduce heat to low and simmer, uncovered, for 20 minutes. Stir occasionally. Serve hot.

Squash with Sweet Veggie Stuffing (Page 69) ➤
Overleaf: Rotini with Red Bell Pepper and Rapini Sauce (Page 117)

PER SERVING

Calcium	Calories	Protein	Fat	Saturated Fat	Carbohydrates	Fiber	Iron	Sodium
276 mg	276	14 g	6 g	0.8 g	46 g	8 g	9 mg	164 mg

Snowy Bean Mashed Potatoes

SERVES 3

2	medium potatoes, peeled and quartered	2
2 tsp	olive oil	10 mL
2 cups	cooked white beans or 1 can (19 oz [540 mL]) rinsed and drained	500 mL
1/4 cup	water	50 mL
3	cloves garlic, minced	3
3/4 tsp	salt	4 mL
1/4 tsp	black pepper	1 mL

1. Boil potatoes for 15 minutes or until soft. Drain.

2. In a bowl, mash potatoes with oil.

3. In a food processor, purée beans, water, garlic, salt and pepper to form a thick paste. Transfer mixture to mashed potatoes in bowl; mix until well combined. Serve hot or at room temperature.

◄ LEMON GINGER STIR-FRY WITH ALMONDS (PAGE 84)

Here's a simple and imaginative way to bring mashed potatoes to your table. By mashing beans and potatoes together, the texture and appeal of traditional mashed potatoes remains, yet the dish gets a calcium boost. These potatoes have a nice garlicky flavor.

TIPS

If serving this dish to children, you may want to eliminate the garlic and pepper.

For a complete meal, serve a generous portion of this dish accompanied by a large, crisp, calcium-rich green salad (see page 168) and a slice of hearty bread.

	PER SERVING							
Calcium	Calories	Protein	Fat	Saturated Fat	Carbohydrates	Fiber	Iron	Sodium
121 mg	273	13 g	4 g	0.5 g	49 g	9 g	5 mg	586 mg

Sweet Cinnamon Potato Boats

SERVES 6

PREHEAT OVEN TO 350° F (180° C)

3	medium-large sweet potatoes	3
1 1/2 cups	cooked soybeans or navy beans or *1 can (14 oz [398 mL])* rinsed and drained	375 mL
1/2 cup	orange juice *(preferably calcium fortified)*	125 mL
1 tsp	olive oil	5 mL
1/2 tsp	cinnamon	2 mL
1/4 tsp	salt	1 mL

1. Pierce sweet potatoes with a fork, wrap in foil and bake about 1 1/2 hours or until soft inside. Allow to cool slightly.

2. With a sharp knife, cut an oval in the long side of each potato. Remove the skin and gently scoop out the flesh into a large bowl, leaving a thickness of about 1/4 inch (5 mm). Set the potato skins aside.

3. In a blender or food processor, combine the cooked beans, orange juice and olive oil; purée until creamy. Add mixture to sweet potatoes and mash together with a potato masher. (For a creamier mixture, purée until smooth using a food processor.) Add cinnamon and salt; stir to mix.

4. Stuff potato-bean mixture into potato skins and bake in preheated oven for 20 minutes.

Here's a fun way to get kids of all ages to eat beans — by disguising them in mashed sweet potatoes, then stuffing them into potato shells. Our family especially enjoys these calcium-rich potato boats because they have a wonderful hint of orange. And because they're so filling, a single potato boat can be a meal in itself!

		PER SERVING						
Calcium	Calories	Protein	Fat	Saturated Fat	Carbohydrates	Fiber	Iron	Sodium
119 mg	267	10 g	5 g	0.7 g	48 g	8 g	3 mg	113 mg

Creamy Sweet Potato Bake

SERVES 4 TO 6 *OR* MAKES 8 POTATO BOATS

PREHEAT OVEN TO 425° F (220° C)

4	medium sweet potatoes	4
1 lb	soft tofu	500 g
2 tbsp	margarine	25 mL
2	large onions, finely diced	2
1 tsp	salt	5 mL
1/4 tsp	black pepper	1 mL
3 tbsp	chopped dill	45 mL
3 tbsp	chopped cilantro	45 mL

1. Pierce sweet potatoes with a fork, wrap in foil and bake about 1 1/2 hours or until soft inside. Allow to cool slightly.

2. With a sharp knife, cut an oval in the long side of the potato. Remove the skin and gently scoop out the flesh into a large bowl, leaving a thickness of 1/4 inch (5 mm). Set the potato skins aside.

3. Add tofu to bowl of sweet potatoes. Mash together until smooth with a potato masher or food processor.

4. In a nonstick skillet, heat margarine over medium heat. Add onions, salt and pepper; sauté until onions have browned. Stir into potato-tofu mixture. Add dill and cilantro; mix well.

5. Stuff potato skins with potato mixture or, if serving alone, transfer potato mixture to a casserole dish. Bake in preheated oven for 10 minutes or until heated through.

Here's a mashed potato dish that everyone can enjoy. You can serve these creamy potatoes as a conventional accompaniment to other foods or as a stuffing for potato skins.

TIP

Any leftover filling can be served on the side or refrigerated and stored for up to 3 days.

				PER SERVING (4)				
Calcium	Calories	Protein	Fat	Saturated Fat	Carbohydrates	Fiber	Iron	Sodium
284 mg	445	13 g	9 g	1.2 g	80 g	10 g	2 mg	623 mg

Layered Mashed Potatoes

SERVES 6 TO 8

PREHEAT OVEN TO 350° F (180° C)
11- BY 9-INCH (2.5 L) GLASS BAKING DISH

One of the most comforting foods are mashed potatoes. In this recipe, white potatoes are layered with sweet potatoes to create a color-ful and delicious casserole that features an abundance of other calcium-rich vegetables. Enjoy a change from conventional mashed potatoes with this fancy and flavorful potato pie.

6	medium white potatoes, peeled and quartered	6
1/4 cup	soy milk (calcium fortified)	50 mL
3	medium-large sweet potatoes, peeled and quartered	3
2 tsp	olive oil	10 mL
1	large onion, finely chopped	1
4 cups	packed finely chopped collard greens, steamed	1 L
1 tsp	salt	5 mL
3 cups	packed finely chopped Chinese lettuce, steamed	750 mL
1/4 cup	finely chopped fresh whole fennel Cilantro leaves	50 mL

1. In a large pot of boiling water, cook white potatoes until soft; drain and transfer to a large bowl. Mash potatoes with soy milk and set aside.

2. In another pot of boiling water, cook sweet potatoes until soft; drain and transfer to another large bowl. Mash and set aside.

3. In a nonstick skillet, heat oil over medium-high heat. Add onion and sauté until browned. Add to mashed white potato, along with collards and salt. Mix well.

4. Add Chinese lettuce and fennel to sweet potatoes; mix well.

5. Make a 1-inch (2.5 cm) layer of white potato mixture in prepared baking dish. Follow with a layer of sweet potato mixture. Finish with white potato layer. Bake in preheated oven for 30 minutes.

6. Sprinkle cilantro over top. Cut into large squares and serve hot.

PER SERVING (6)

Calcium	Calories	Protein	Fat	Saturated Fat	Carbohydrates	Fiber	Iron	Sodium
208 mg	345	7 g	2 g	0.3 g	77 g	10 g	2 mg	427 mg

Faux French Fries

MAKES ABOUT 40 FRIES, SERVING 3

PREHEAT OVEN TO 400° F (200° C)
2 NONSTICK BAKING SHEETS, LIGHTLY OILED

1	small butternut or acorn squash	1
2	medium sweet potatoes	2
	Vegetable oil	

1. Peel vegetables and slice them into the shape of oversized french fries, about 1 inch (2.5 cm) thick. Place each type of fries in single layer on separate prepared baking sheet. Bake for 25 to 30 minutes, turn, then bake another 25 to 30 minutes or until edges become browned. Check for doneness; if you like your chips crispier, increase the baking time accordingly.

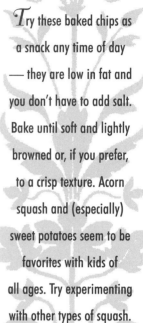

Try these baked chips as a snack any time of day — they are low in fat and you don't have to add salt. Bake until soft and lightly browned or, if you prefer, to a crisp texture. Acorn squash and (especially) sweet potatoes seem to be favorites with kids of all ages. Try experimenting with other types of squash.

TIP

Prior to baking, add great flavor by brushing fries with a little oil then tossing with garlic salt, pepper or dried herbs such as parsley or cilantro.

PER SERVING

Calcium	Calories	Protein	Fat	Saturated Fat	Carbohydrates	Fiber	Iron	Sodium
140 mg	282	5 g	0.4 g	0.1 g	69 g	9 g	2 mg	28 mg

Burgers & Patties

Burgers and patties of all types, shapes and sizes are as familiar as they are versatile. Here we provide calcium-rich recipes that are easy to prepare and can be made in large batches to be frozen and served at a later date.

These burgers and patties can be served in pita bread or whole-wheat buns and stuffed with all the trimmings — lettuce, onions, tomatoes and other favorites. Another idea: take a couple of previously cooked burgers from the freezer, thaw and break them up, then mix with some store-bought tomato sauce. Heat, then toss the sauce with a small amount of cooked pasta and you'll have an almost instant calcium-rich meal.

In our experience, children — especially toddlers — enjoy burgers because they can pick them up and feed themselves. When introducing a child to the choices here, serve plain — either warm or at room temperature.

These recipes can also be a great fast food. Take a pre-cooked burger, patty or ball directly from the freezer and place it into a zip-lock bag for a quick snack when you're on the go.

Great Grain Burgers

1 tbsp	olive oil	15 mL
1 cup	finely chopped broccoli	250 mL
1 cup	finely chopped onions	250 mL
3	stalks celery, finely chopped	3
1 tsp	garlic salt	5 mL
1 tsp	ground cumin	5 mL
1 cup	vegetable stock	250 mL
3 1/2 cups	finely chopped collard greens	875 mL
1 cup	cooked brown rice (see page 20 for technique)	250 mL
1 cup	cooked quinoa (see page 20 for technique)	250 mL
8 oz	firm tofu, crumbled	250 g
1 cup	rolled oats	250 mL
1/4 cup	ground almonds	50 mL
3 tsp	tamari sauce	15 mL
1 tsp	salt	5 mL
1 tsp	black pepper	5 mL
1/2 cup	bread crumbs	125 mL

TIP

When forming these burgers, use wet hands to keep from sticking.

1. In a large nonstick skillet, heat oil over medium heat. Add broccoli, onions and celery; sauté 5 minutes. Add garlic salt, cumin and stock. Reduce heat and simmer 15 minutes. Add collards; mix well, cover and simmer for another 7 minutes, stirring occasionally. Transfer mixture to a large bowl and let cool.

2. Add to cooled mixture the cooked brown rice and quinoa, tofu, oats, almonds, tamari sauce, salt and pepper; mix well. Transfer 2 cups (500 mL) of the mixture to a food processor; process until it becomes thick and sticky. Return processed mixture to bowl and combine. Add bread crumbs and mix well. Refrigerate 30 minutes.

PER BURGER								
Calcium	Calories	Protein	Fat	Saturated Fat	Carbohydrates	Fiber	Iron	Sodium
91 mg	126	6 g	4 g	0.6 g	18 g	2 g	3 mg	385 mg

3. With your hands, form burgers about 1/4 inch (5 mm) thick. In a lightly oiled nonstick skillet, fry burgers until browned on both sides.

Hearty Soybean Burgers

MAKES ABOUT 10 BURGERS

PREHEAT OVEN TO 350° F (180° C)
NONSTICK BAKING SHEET, LIGHTLY OILED

These hearty calcium-rich burgers make for a satisfying meal, especially served on a bun with all your favorite condiments. They're also delicious with a savory sauce poured on top. Try one of our calcium-rich tomato sauces (see Pasta section, page 115, for ideas).

3 cups	cooked soybeans or *2 cans (each 14 oz [398 mL])* *rinsed and drained*	750 mL
2	large potatoes, peeled and quartered	2
1 1/2 tbsp	olive oil	20 mL
2	small onions, finely chopped	2
3	carrots, grated	3
1	stalk celery, finely chopped	1
8 oz	extra firm tofu, crumbled	250 g
1/2 cup	bread crumbs	125 mL
1/3 cup	ground almonds	75 mL
3 tsp	prepared mustard	15 mL
2 tsp	salt	10 mL
2 tsp	dried basil	10 mL
1 1/2 tsp	dried oregano	7 mL
1 tsp	garlic salt	5 mL
1/2 tsp	dried thyme	2 mL

1. In a large bowl, mash cooked soybeans.

2. In a large pot of boiling water, cook potatoes until soft; drain, mash and add beans. Mix well.

3. In a nonstick skillet, heat oil over medium heat. Add onions, carrots and celery; sauté until soft. Add to bean-potato mixture along with tofu, bread crumbs, ground almonds, mustard, salt, basil, oregano, garlic salt and thyme; mix well.

4. With your hands, form burgers and place on baking sheet. Bake 20 minutes; turn over, then bake another 20 minutes.

PER BURGER								
Calcium	Calories	Protein	Fat	Saturated Fat	Carbohydrates	Fiber	Iron	Sodium
142 mg	218	14 g	9 g	1.2 g	22 g	5 g	3 mg	659 mg

Chickpea Bean Burgers

MAKES ABOUT 10 BURGERS

PREHEAT OVEN TO 350° F (180° C)
NONSTICK BAKING SHEET, LIGHTLY OILED

2 cups	cooked white or navy beans or 1 can (19 oz [540 mL]) rinsed and drained	500 mL
1 cup	cooked chickpeas or 1 can (14 oz [398 mL]) rinsed and drained	250 mL
1 1/2 tbsp	olive oil	20 mL
2	stalks celery, finely chopped	2
1	medium onion, finely chopped	1
1/2 cup	tomato paste	125 mL
1/2 cup	chopped dill	125 mL
1/2 cup	bread crumbs	125 mL
1 tsp	salt	5 mL

These delicious, versatile bean burgers are crunchy on the outside yet soft on the inside. They can be served alone, in a bun with condiments, or with your favorite sauce on top. Or try them in a sandwich with shredded lettuce and chopped tomato along with our MOCK MAYONNAISE (see recipe, page 35).

1. In a large bowl, mash white beans and chickpeas together. Set aside.

2. In a nonstick skillet, heat oil over medium-high heat. Add celery and onion; sauté until soft. Remove from heat and add to mashed beans along with tomato paste, dill, bread crumbs and salt; mix well.

3. With your hands, form into 2 1/2-inch (6 cm) patties; place on baking sheet. Bake for 25 minutes; turn over and bake another 20 minutes.

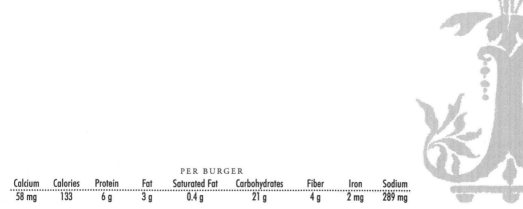

| | | | PER BURGER | | | | | | |
Calcium	Calories	Protein	Fat	Saturated Fat	Carbohydrates	Fiber	Iron	Sodium
58 mg	133	6 g	3 g	0.4 g	21 g	4 g	2 mg	289 mg

Millet and Kale Patties

MAKES ABOUT 10 PATTIES

PREHEAT OVEN TO 350° F (180° C)
NONSTICK BAKING SHEET, LIGHTLY OILED

These calcium-rich patties are so flavorful you can enjoy them just as they are. However, for added interest, you can melt a slice of soy cheese on top and serve in a bun or pita.

1 1/2 tbsp	canola oil	20 mL
3	large onions, finely diced	3
4	cloves garlic, minced	4
2 cups	finely chopped bok choy	500 mL
5 cups	packed finely chopped kale	1.25 L
2	carrots, finely diced	2
1	parsnip, finely diced	1
1/4 cup	soy sauce	50 mL
1 tsp	dried basil	5 mL
1 tbsp	chopped parsley	15 mL
1 tsp	salt	5 mL
1/2 tsp	paprika	2 mL
3 cups	cooked millet (see Tip, at left)	750 mL
1/4 cup	almond butter	50 mL
3 tbsp	ketchup	45 mL
1/2 cup	amaranth flour or soy flour	125 mL
1/2 cup	whole wheat flour	125 mL

TIPS

Try these patties fried: dredge lightly in flour and fry in a well oiled nonstick skillet until browned on both sides.

To cook millet: With a strainer, rinse 1 cup (250 mL) millet; drain. Add to pot with 2 cups (500 mL) water; bring to a boil. Reduce heat and simmer 30 minutes.

1. In a large nonstick skillet with a lid, heat oil over medium-high heat. Add onions, garlic and bok choy; sauté 4 minutes. Add kale, carrots, parsnip and soy sauce; stir to combine. Reduce heat, cover and simmer, stirring occasionally, for 20 minutes or until vegetables are soft. Add basil, parsley, salt and paprika; mix well. Remove from heat.

2. Transfer vegetable mixture to a large bowl. Add cooked millet, almond butter and ketchup; mix well. Gradually stir in amaranth and whole wheat flour; mix well to a thick consistency. With your hands, form into 4-inch (10 cm) patties and place on a baking sheet. Bake 15 minutes on one side then flip over and bake an additional 20 minutes or until browned.

PER PATTY

Calcium	Calories	Protein	Fat	Saturated Fat	Carbohydrates	Fiber	Iron	Sodium
142 mg	236	8 g	7 g	0.8 g	38 g	6 g	3 mg	747 mg

Savory Bean Patties

2 1/2 cups	cooked white or black beans or 2 cans (each 14 oz [398 mL]) rinsed and drained	625 mL
2 cups	bread crumbs	500 mL
1 cup	finely chopped bok choy	250 mL
2	packages onion soup mix	2
3/4 cup	finely chopped almonds	175 mL
1/2 cup	finely chopped celery	125 mL
1/2 cup	finely chopped green onions	125 mL
3 tbsp	tomato paste	45 mL
2 tbsp	blackstrap molasses	25 mL
3	large cloves garlic, minced	3
1 tbsp	soy sauce	15 mL
1 tbsp	lemon juice	15 mL
2 tsp	dry mustard	10 mL

1. In a large bowl, mash white beans to a lumpy consistency. Add remaining ingredients and mix well.

2. With your hands, form mixture into patties. In a large, lightly oiled, nonstick skillet over medium heat, fry each side until slightly crusty and browned.

These bean patties are an enjoyable and filling main dish. They are thick and soft with a bit of a crunch from the chopped almonds, celery and onions. Try them in a bun with your favorite condiments or topped with a high-calcium sauce (see Pasta section, page 115, for ideas), accompanied by a side salad.

			PER PATTY					
Calcium	Calories	Protein	Fat	Saturated Fat	Carbohydrates	Fiber	Iron	Sodium
103 mg	165	7 g	4 g	0.5 g	26 g	3 g	3 mg	549 mg

Sweet Potato Mini Patties

MAKES ABOUT 40 PATTIES

PREHEAT OVEN TO 350° F (180° C)

We love these sweet, soft and colorful orange patties because they can be served as a main dish, side dish or eaten as a snack. They are small and tasty, so chances are you'll want more than just one.

1/2 cup	cooked chickpeas or *canned, rinsed and drained*	125 mL
4	medium sweet potatoes	4
1/2 tbsp	olive oil	7 mL
1	onion, finely chopped	1
2 cups	finely chopped Chinese lettuce	500 mL
1	red bell pepper, seeded and finely chopped	1
4	cloves garlic, minced	4
2 tsp	salt	10 mL
1/4 tsp	dried oregano	1 mL
2 cups	rolled oats	500 mL
1/2 cup	wheat germ	125 mL
1/3 cup	whole wheat flour	75 mL
3/4 cup	bread crumbs	175 mL

1. In a bowl using a potato masher or in a food processor, mash chickpeas to a rough paste. Set bowl aside.

2. Pierce sweet potatoes with a fork; wrap in foil and bake in a preheated oven for 1 1/2 hours or until soft. Remove skins and mash together with cooked chickpeas. Set mixture aside.

3. In a nonstick skillet with a lid, heat oil over medium heat. Add onion and Chinese lettuce; sauté for 3 minutes. Add red pepper, garlic, salt and oregano. Cover and cook 4 minutes or until vegetables are soft. Transfer to sweet potato-chickpea mixture along with rolled oats, wheat germ and flour. Mix until thick and well combined. Let cool.

4. With slightly wet hands, form mixture into small patties, each about 1 1/2-inch (3 cm) round and flattened. Dredge in bread crumbs and, in a well oiled nonstick skillet, fry on both sides until browned.

				PER PATTY				
Calcium	Calories	Protein	Fat	Saturated Fat	Carbohydrates	Fiber	Iron	Sodium
19 mg	70	2 g	1 g	0.1 g	14 g	2 g	1 mg	134 mg

Potato Veggie Patties

MAKES 25 TO 30 SMALL PATTIES

6 cups	*finely chopped beet greens*	1.5 L
1	*small sweet potato, peeled and grated, excess liquid drained*	1
5	*medium white potatoes, peeled and grated, excess liquid drained*	5
1	*onion, finely chopped*	1
1	*stalk celery, finely chopped*	1
1/4 cup	*chopped dill*	50 mL
1 1/2 cups	*bread crumbs*	375 mL
1/4 cup	*all-purpose flour*	50 mL
1 1/2 tsp	*salt*	7 mL
1 tsp	*baking powder*	5 mL
1/2 tsp	*garlic salt*	2 mL

1. In a large pot, steam beet greens for 3 to 4 minutes or until wilted. Remove from heat, cool, squeeze out excess water and transfer to a large bowl. Add grated potatoes, onion, celery, dill, bread crumbs, flour, salt, baking powder and garlic salt; mix well.

2. With slightly damp hands, form mixture into small, thin patties. In a well oiled nonstick skillet over medium heat, fry patties until brown and crisp on both sides. Transfer to paper towel to absorb excess oil. Serve hot.

Calcium-rich beet greens make these patties wonderfully nutritious. Once cooked, the red rib of the beet greens adds an interesting festive appearance. Serve these tasty veggie patties alone or in a bun or pita with lettuce and tomatoes. You will need 3 or 4 of them to stuff a pita or bun.

TIP

When grating potatoes it is best to use a hand grater; this helps the patties hold together better.

	Calcium	Calories	Protein	Fat	Saturated Fat	Carbohydrates	Fiber	Iron	Sodium
PER PATTY	32 mg	66	2 g	0 g	0.1 g	14 g	1 g	1 mg	249 mg

Broccoli Potato Mini Patties

MAKES ABOUT 35 PATTIES

These mini patties are a tasty and creative way to serve broccoli, which is blended with potato and seasonings. Toddlers love to pick them up in their tiny hands and feed themselves. These patties also work well as a main meal or snack. (You will need to eat at least 4 of them to fill yourself up.)

5 cups	chopped broccoli	1.25 L
3	small potatoes, peeled and quartered	3
1/2 tbsp	olive oil	7 mL
1	medium onion, finely chopped	1
1 1/2	stalks celery, finely chopped	1 1/2
2/3 cup	bread crumbs, divided	150 mL
3 tbsp	finely chopped cilantro	45 mL
2 tsp	lemon juice	10 mL
1 tsp	salt	5 mL
1/2 tsp	black pepper	2 mL

1. In a large pot, steam broccoli until soft. Transfer to a bowl and set aside.

2. In a large pot of boiling water, cook potatoes until soft. Drain and add to steamed broccoli. Mash together.

3. In a saucepan heat oil over medium heat. Add onion and celery; cook until tender. Add to broccoli-potato mixture. Mix well, cover and refrigerate for 30 minutes.

4. Add 2 tbsp (25 mL) of the bread crumbs, the cilantro, lemon juice, salt and pepper; mix well. With your hands, form into 1-inch (2.5 cm) balls, then flatten to form mini patties. Dredge lightly to coat in remaining bread crumbs. In a well oiled nonstick skillet over medium heat, fry patties for about 5 minutes per side or until golden brown.

TIP

If you prefer, bake the patties on a lightly greased baking sheet in a preheated 350° F (180° C) oven for 10 minutes; turn patties over and bake another 15 minutes.

				PER PATTY				
Calcium	Calories	Protein	Fat	Saturated Fat	Carbohydrates	Fiber	Iron	Sodium
10 mg	22	1 g	0.4 g	0.1 g	4 g	1 g	0.2 mg	86 mg

Nutty Falafel Balls

MAKES ABOUT 35 BALLS

PREHEAT OVEN TO 350° F (180° C)
NONSTICK BAKING SHEET, LIGHTLY GREASED

3 cups	cooked chickpeas or 2 cans (each 14 oz [398 mL]) rinsed and drained	750 mL
1 cup	finely chopped brazil nuts	250 mL
1 cup	finely chopped bok choy	250 mL
1 1/2 tbsp	olive oil	20 mL
1	onion, finely chopped	1
1/4 cup	whole wheat flour	50 mL
4	cloves garlic, minced	4
1 tsp	ground cumin	5 mL
2 tsp	ground coriander	10 mL
1 tsp	salt	5 mL
1/2 tsp	black pepper	2 mL

These nutritious falafel balls can be baked or fried. They're a great addition to any meal. Children love the taste, texture and shape. Enjoy them as a healthy snack or sandwich stuffer.

1. In a large bowl, mash cooked chickpeas. Add brazil nuts and bok choy; mix until combined. Transfer in 2 batches to a food processor, processing each until mixture achieves a sticky texture. Return to bowl.

2. Add olive oil, onion, flour, garlic, cumin, coriander, salt and pepper to chickpea mixture; combine until well mixed. With your hands, form mixture into 1 1/4-inch (3 cm) balls and place on baking sheet. Bake 40 minutes, turning balls over at the halfway point.

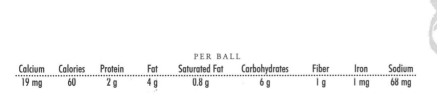

	PER BALL							
Calcium	Calories	Protein	Fat	Saturated Fat	Carbohydrates	Fiber	Iron	Sodium
19 mg	60	2 g	4 g	0.8 g	6 g	1 g	1 mg	68 mg

Kale and Soy Cheese Balls

MAKES ABOUT 15 BALLS

PREHEAT OVEN TO 325° F (160° C)
NONSTICK BAKING SHEET, LIGHTLY GREASED

3 cups	finely chopped kale	750 mL
2 cups	bread crumbs (whole wheat or regular)	500 mL
3/4 cup	vegetable stock	175 mL
1/2 cup	grated soy cheese	125 mL
1/4 cup	finely chopped onions	50 mL
1 tbsp	olive oil	15 mL
1/2 tsp	soy sauce	2 mL
1/2 tsp	garlic salt	2 mL

1. In a large pot, steam kale until soft; allow to cool. Squeeze out excess moisture and transfer to a large bowl. Add remaining ingredients and mix well.

2. With your hands, shape mixture into small balls about 1 1/4 inches (3 cm) round. Place on baking sheet. Bake 35 to 40 minutes, turning them over after first 15 minutes.

Slightly firm on the outside and soft on the inside, these kale and cheese balls are a delicious addition to any meal. Serve with a little ketchup or tomato sauce, or try spooning one of our calcium-rich pasta sauces over them. (See Pasta section, page 115). Toddlers may enjoy eating them plain.

	PER BALL							
Calcium	Calories	Protein	Fat	Saturated Fat	Carbohydrates	Fiber	Iron	Sodium
53 mg	86	3 g	2 g	0.4 g	13 g	2 g	1 mg	202 mg

Pasta

Pasta is the all-time favorite for many people. From thin noodles to thick, enriched or flavored, packaged or fresh, the variety of pasta is endless. So, too, is the variety of techniques for preparing pasta.

In this chapter, you'll find a selection of calcium-rich pasta recipes that not only taste great but also add color, texture and zest to your meals.

Try using the pasta sauces in this chapter with a host of other foods, pouring them over baked potatoes, grains or vegetables. It's a great way to add calcium — and flavor.

Pasta sauce can be frozen in small containers and used at your convenience. When trying these recipes you can always substitute one pasta for another of your favorites.

Lazy Day Lasagna

SERVES 8 OR 9

PREHEAT OVEN TO 350° F (180° C)
13- BY 9-INCH (3 L) BAKING DISH

This lasagna tastes and looks like the cheese-laden original, but with delicious layers of dark greens and beans. It's a healthy, calcium-rich alternative to conventional lasagna.

9 to 12	lasagna noodles	9 to 12
2 cups	cooked white or navy beans or *1 can (19 oz [540 mL])* rinsed and drained	500 mL
8 oz	firm tofu, crumbled	250 g
4 cups	packed finely chopped collard greens, lightly steamed	1 L
2	cans pasta sauce (each 19 oz [540 mL]), store-bought variety	2
3	cloves garlic, minced	3
3 1/2 cups	shredded soy cheese	875 mL

1. In a pot of boiling salted water, cook lasagna noodles for 8 to 10 minutes or until tender but firm; drain. Rinse under cold running water. Drain and set aside.

2. In a large bowl, mash cooked beans. Add tofu; mix well. Stir in collards, pasta sauce and garlic.

3. Pour tomato mixture to coat bottom of baking dish. Layer with noodles, then sauce, then soy cheese. Repeat layering until ingredients are used up. The top layer should be soy cheese. Bake uncovered in preheated oven for 45 minutes.

TIP

For added flavor and texture, choose a pasta sauce that has added ingredients, such as mushrooms, red peppers or olives.

PER SERVING (8)

Calcium	Calories	Protein	Fat	Saturated Fat	Carbohydrates	Fiber	Iron	Sodium
420 mg	487	30 g	15 g	2.0 g	59 g	8 g	7 mg	792 mg

Rotini with Red Bell Pepper and Rapini Sauce

SERVES 2 TO 4

MAKES 2 1/2 CUPS SAUCE

PREHEAT BROILER

10 oz	rotini	300 g
	Sauce	
3	red bell peppers	3
1 tbsp	olive oil	15 mL
4	cloves garlic, minced	4
1 cup	finely chopped rapini	250 mL
2 cups	chopped bok choy	500 mL
1/3 cup	vegetable stock	75 mL
1 tbsp	chopped basil	15 mL
2 tbsp	white wine (optional)	25 mL
1 tsp	salt	5 mL
1/2 tsp	black pepper	2 mL

1. In a large pot of boiling water, cook pasta until tender but firm. Drain, cover and set aside.

2. Place bell peppers under broiler, turning until all sides are charred. Remove from oven and let cool. Peel off skin and remove seeds. Chop peppers and set aside.

3. In a large nonstick saucepan, heat oil over medium-high heat. Add garlic and sauté for 30 seconds. Add rapini and bok choy; sauté 5 minutes. Add chopped peppers, vegetable stock, basil, wine, salt and pepper; stir together and cook 5 minutes.

4. In a food processor or blender, process vegetable mixture until slightly chunky. Return to saucepan and simmer 2 minutes. Serve hot over rotini.

We often serve this light and colorful sauce to our guests — it's one of our favorites! Not only do the rapini and bok choy offer a great calcium kick, but you can prepare this colorful sauce in a variety of ways — purée it until smooth or leave it slightly chunky and serve it over steamed vegetables, potatoes or grains.

PER SERVING (2)

Calcium	Calories	Protein	Fat	Saturated Fat	Carbohydrates	Fiber	Iron	Sodium
197 mg	666	22 g	10 g	1.4 g	122 g	11 g	4 mg	1298 mg

Rigatoni Vegetable Pasta

SERVES 4 TO 6

MAKES 5 CUPS (1.25 L) SAUCE

1 lb	rigatoni	500 g

Sauce

2 1/2 cups	tomato sauce	625 mL
2 tbsp	tomato paste	25 mL
2 cups	chopped bok choy, steamed	500 mL
1 cup	chopped broccoli, steamed	250 mL
1 cup	chopped green beans, steamed	250 mL
2	carrots, chopped and steamed	2
1	onion, chopped and steamed	1
8 oz	regular or soft tofu, crumbled	250 g
2 tbsp	chopped parsley	25 mL
1/2 tbsp	olive oil	5 mL
3	cloves garlic, minced	3
1 tsp	dried oregano	5 mL
1	bay leaf	1
1/2 tsp	salt	2 mL
1/4 tsp	black pepper	1 mL

1. In a large pot of boiling water, cook pasta until tender but firm. Drain and set aside in serving bowl.

2. In a food processor, blend tomato sauce, tomato paste, bok choy, broccoli, green beans, carrots and onion until smooth. Transfer to a large pot; simmer over low heat until heated through. Stir in tofu, parsley, oil, garlic, oregano, bay leaf, salt and pepper. Cover pot and simmer for 15 minutes.

3. Remove bay leaf and pour sauce over rigatoni. Serve hot.

Loaded with nutritious vegetables, the sauce in this recipe is one that everyone can enjoy. It's calcium-rich and flavorful, whether used in a lasagna recipe or spooned over pasta. In fact, this sauce is so good, we often eat it without the pasta!

TIP

Instead of crumbling the tofu, try slicing it into small squares (or scoop with a melon baller for small round pieces) and dropping them into your sauce. The tofu will absorb the flavor of the sauce and create a chunkier topping for pasta or grains.

PER SERVING (4)								
Calcium	Calories	Protein	Fat	Saturated Fat	Carbohydrates	Fiber	Iron	Sodium
188 mg	583	24 g	7 g	1.0 g	109 g	11 g	7 mg	1267 mg

Angel Hair Pasta with Creamy Tofu-Tomato Sauce

SERVES 4 TO 6

MAKES 4 CUPS (1 L) SAUCE

12 oz	angel hair pasta	375 g
	Sauce	
1 tsp	olive oil	5 mL
1	onion, chopped	1
3	large cloves garlic, finely chopped	3
12 oz	soft tofu	375 g
2 1/2 cups	tomato sauce	625 mL
2 cups	packed finely chopped kale, steamed	500 mL
2 1/2 cups	finely chopped broccoli, steamed	625 mL
1/2 tbsp	tamari sauce	7 mL
1/2 tsp	salt	2 mL

1. In a large pot of boiling water, cook pasta until tender but firm. Drain and set aside.

2. In a small saucepan, heat oil over medium heat. Add onion and garlic; sauté until softened. Set aside.

3. In a food processor or blender, purée tofu and tomato sauce until creamy and smooth. Transfer mixture to medium-sized pot and heat 3 to 4 minutes. Stir in kale, broccoli, sautéed onion and garlic, along with tamari and salt. Cover pot and simmer over low heat 7 to 8 minutes, stirring occasionally.

4. Pour sauce over pasta and serve.

The rich, pink-colored sauce in this recipe is particularly enjoyable with angel hair pasta. Best of all, the tofu, kale and broccoli make a powerful calcium combination. Try sprinkling some chopped fresh basil over top for an appealing garnish.

CALCIUM REPORT

In a 1974 study, the Alaskan Inuit were found to have the highest known intake of calcium and animal protein. Yet they were found to be suffering from some of the worst known rates of osteoporosis.

PER SERVING (4)

Calcium	Calories	Protein	Fat	Saturated Fat	Carbohydrates	Fiber	Iron	Sodium
257 mg	474	21 g	6 g	0.8 g	86 g	8 g	3 mg	1356 mg

Penne with Fennel Chickpea Tomato Sauce

SERVES 6

MAKES 6 CUPS (1.5 L) SAUCE

1 lb	*penne*	*500 g*

Sauce

1 tbsp	*olive oil*	*15 mL*
1	*small onion, chopped*	*1*
1	*small red bell pepper, chopped*	*1*
2	*cloves garlic, finely chopped*	*2*
1/4 tsp	*chili flakes*	*1 mL*
8 oz	*regular or soft tofu, crumbled*	*250 g*
4 oz	*soft tofu, crumbled*	*125 g*
1 1/2 cups	*cooked chickpeas* *or 2 cans (each 14 oz [398 mL])* *drained and rinsed*	*375 mL*
1	*small tomato, chopped*	*1*
2 1/2 cups	*Italian-style tomato sauce*	*625 mL*
1/2 cup	*vegetable stock*	*125 mL*
1/2 cup	*chopped fresh fennel*	*125 mL*
1/4 cup	*chopped parsley*	*50 mL*
1/2 tbsp	*dried oregano*	*7 mL*
1/2 tbsp	*dried basil*	*7 mL*
1 tsp	*salt*	*5 mL*
1 cup	*sun-dried tomatoes,* *drained and finely chopped*	*250 mL*
4 oz	*firm tofu, cut into 1/4-inch* *(5 mm) cubes*	*125 g*

CALCIUM TIP

You can try spooning this sauce over quinoa to increase the calcium of your meal.

1. In a large pot of boiling water, cook pasta until tender but firm. Drain and set aside.

		PER SERVING							
Calcium	Calories	Protein	Fat	Saturated Fat	Carbohydrates	Fiber	Iron	Sodium
186 mg	524	23 g	12 g	1.6 g	85 g	9 g	7 mg	1122 mg

2. In a large saucepan with a lid, heat oil over medium heat. Add onion, bell pepper, garlic and chili flakes; cook for 5 minutes. Add tofu, chickpeas, tomato, tomato sauce, stock, fennel, parsley, oregano, basil and salt. Simmer, covered, for 10 minutes. Remove from heat and allow flavors to develop for another 10 minutes.

3. In a food processor or blender, purée sauce until smooth. Return to saucepan; stir in sun-dried tomatoes and cubed tofu. Serve hot over pasta.

Chunky Fettuccine with Fresh Herbs

SERVES 4

MAKES 3 1/2 CUPS (875 mL) SAUCE

Light-colored and creamy, the sauce used here is also delicious poured over linguine or spaghetti. For a fabulous change, spoon sauce over baked potatoes or increase the calcium by spooning the sauce over cooked quinoa.

TIPS

Leftover squash can be frozen and used at a later date. It also makes a great addition to your favorite soup.

If using the sauce with baked potatoes, garnish with some grated soy cheese over top.

12 oz	fettuccine	375 g
	Sauce	
1	acorn squash, peeled and cubed, divided	1
1 1/2 cups	soy milk (calcium fortified), divided	375 mL
1/2 tbsp	olive oil	7 mL
2 cups	finely chopped bok choy	500 mL
1	medium onion, finely chopped	1
1	stalk celery, finely chopped	1
2	large cloves garlic, minced	2
1/4 cup	water	50 mL
3 tbsp	chopped parsley	45 mL
2 tbsp	chopped basil	25 mL
1/2 tsp	dried ground rosemary	2 mL
1 tsp	salt	5 mL
1/8 tsp	black pepper	0.5 mL
	Chopped basil	

1. In a large pot of boiling water, cook pasta until tender but firm. Drain and set aside in a large serving bowl.

2. In a pot of boiling water, cook squash until soft; drain. Place 2 cups (500 mL) of the squash in blender or food processor. (Save and freeze any remaining squash cubes for another use.) Add 1/2 cup (125 mL) of the soy milk and purée until smooth. Set aside.

3. In a large pot, heat oil over medium heat. Add bok choy, onion, celery and garlic; sauté for 3 minutes. Add water, parsley, basil, rosemary, salt and pepper; sauté 5 minutes. Add puréed squash; mix and cook over low heat for 5 minutes. Stir in remaining 1 cup (250 mL) of soy milk; simmer 5 minutes.

4. Pour sauce over pasta; toss. Garnish with basil and serve.

PER SERVING

Calcium	Calories	Protein	Fat	Saturated Fat	Carbohydrates	Fiber	Iron	Sodium
215 mg	422	15 g	5 g	0.7 g	80 g	9 g	3 mg	614 mg

Desserts

Here you'll find familiar desserts such as cookies, cakes, pies, frozen treats and candied nuts. Created to satisfy the most discerning sweet tooth, these desserts provide calcium in tasty and fun ways your entire family will enjoy. Children in particular will like these sensational goodies and you can be assured that they're getting calcium-rich goodness with each bite.

Almond Peanut Butter Cookies

MAKES ABOUT 30 COOKIES

PREHEAT OVEN TO 350° F (180° C)
BAKING SHEET, LIGHTLY GREASED

These delicious cookies have a nutty peanut butter flavor and are wonderfully soft and chewy. Whenever we make them, our daughter and her friends quickly gobble them up! You can find almond butter in most health food stores.

3 tbsp	finely chopped dried figs	45 mL
3/4 cup	almond butter	175 mL
2/3 cup	peanut butter	150 mL
1 cup	granulated sugar	250 mL
1/4 cup	ground almonds	50 mL
1/4 cup	maple syrup	50 mL
1 cup	soy milk (calcium fortified)	250 mL
1 tsp	vanilla	5 mL
1 cup	amaranth flour	250 mL
1 cup	all-purpose flour	250 mL
1 tsp	baking soda	5 mL
1/2 tsp	baking powder	2 mL
1/2 tsp	salt	2 mL
3/4 cup	peanut butter chips (optional)	175 mL
30	whole or slivered almonds	30

1. In a large bowl, combine figs, almond butter, peanut butter, sugar, ground almonds, maple syrup, soy milk and vanilla; mix well. Transfer to food processor and blend until smooth. Transfer mixture to a large bowl.

2. In a separate bowl, sift together amaranth flour, all-purpose flour, baking soda, baking powder and salt. Add to almond butter mixture; mix well. Stir in peanut butter chips.

3. With a tablespoon, drop cookie dough onto sheet, making approximately 30 cookies. Decorate each cookie with an almond placed on top.

4. Bake 20 minutes or until edges turn brown. Allow cookies to cool before serving.

			PER COOKIE					
Calcium	Calories	Protein	Fat	Saturated Fat	Carbohydrates	Fiber	Iron	Sodium
46 mg	153	4 g	8 g	1.1 g	18 g	2 g	1 mg	112 mg

Snappy Ginger Cookies

MAKES ABOUT 40 COOKIES

PREHEAT OVEN TO 300° F (150° C)
BAKING SHEET, LIGHTLY GREASED

1/2 cup	blackstrap molasses	125 mL
2 tbsp	maple syrup	25 mL
2 tbsp	canola oil	25 mL
1 1/2 tsp	vanilla	7 mL
1 tsp	soy milk (calcium fortified)	5 mL
1 1/4 cups	all-purpose flour	300 mL
1/2 cup	amaranth flour	125 mL
1/4 cup	ground almonds	50 mL
1 1/4 tsp	ground ginger	6 mL
1/2 tsp	baking soda	2 mL
1/2 tsp	baking powder	2 mL
1/2 tsp	cinnamon	2 mL

These great-tasting cookies combine the spicy flavor of ginger with the sweetness of molasses. A favorite with kids and adults alike, these cookies are also great teething biscuits for babies.

1. In a large bowl, combine molasses, maple syrup, oil, vanilla and soy milk; mix well. Set aside.

2. In a separate bowl, sift together all-purpose flour, amaranth flour, almonds, ginger, baking soda, baking powder and cinnamon. Slowly beat into molasses mixture until combined. Knead dough until it firms. Refrigerate for 45 minutes.

3. With a lightly oiled rolling pin, roll out dough on a sheet of lightly greased wax paper. (The thinner you roll the dough, the crispier the cookie will be.) With a cookie cutter or knife, cut cookie shapes. Place cookies on baking sheet.

4. Bake in preheated oven for 10 minutes. Allow cookies to cool before serving.

			PER COOKIE					
Calcium	Calories	Protein	Fat	Saturated Fat	Carbohydrates	Fiber	Iron	Sodium
42 mg	42	1 g	1 g	0.1 g	7 g	0.4 g	1 mg	22 mg

Chocolate Chip Sesame Cookies

MAKES ABOUT 25 COOKIES

PREHEAT OVEN TO 350° F (180° C)
BAKING SHEET, LIGHTLY GREASED

For a mouth-watering treat, try these traditional cookies. They're lightly sweetened and chewy — the perfect accompaniment to a cup of hot chocolate or afternoon tea.

1 cup	granulated sugar	250 mL
3/4 cup	finely chopped almonds	175 mL
1/2 cup	maple syrup	125 mL
1/2 cup	soy milk (calcium fortified)	125 mL
3 tbsp	sesame seeds	45 mL
2 tbsp	canola oil	25 mL
1 tsp	vanilla	5 mL
1 1/4 cups	all-purpose flour	300 mL
1/2 cup	amaranth flour or soy flour	125 mL
1 tsp	baking powder	5 mL
1/4 tsp	salt	1 mL
1 1/2 cups	chocolate chips (dairy-free variety)	375 mL

TIP

Great tasting dairy-free chocolate chips can be purchased at health food stores.

1. In a large bowl, combine sugar, almonds, maple syrup, soy milk, sesame seeds, oil and vanilla. Beat until well mixed.

2. In a separate bowl, sift together all-purpose flour, amaranth flour, baking powder and salt. Gradually beat into sugar mixture until well mixed. Stir in chocolate chips.

3. With a tablespoon, drop heaping spoonsful (15 to 20 mL) of batter onto baking sheet. Bake for 15 minutes or until cookies brown on bottom. Allow cookies to cool before serving.

PER COOKIE

Calcium	Calories	Protein	Fat	Saturated Fat	Carbohydrates	Fiber	Iron	Sodium
33 mg	171	3 g	8 g	2.6 g	25 g	2 g	1 mg	38 mg

Coconut Almond Cookies

MAKES ABOUT 24 COOKIES

PREHEAT OVEN TO 375° F (190° C)
BAKING SHEET, LIGHTLY GREASED

6 oz	soft tofu	175 g
1/4 cup	maple syrup	50 mL
2 tbsp	margarine	25 mL
1 tsp	vanilla	5 mL
1 cup	all-purpose flour	250 mL
1 cup	shredded coconut	250 mL
3/4 cup	granulated sugar	175 mL
3/4 cup	ground almonds	175 mL
1/4 cup	graham cracker crumbs	50 mL
1/4 cup	carob powder	50 mL
1 tsp	baking powder	5 mL
1/4 tsp	salt	1 mL

If you love coconut, these cookies will definitely satisfy your craving. They're soft and chewy — with carob flour, tofu and almonds to boost the calcium.

1. In a large bowl with an electric mixer (or in a food processor or blender), combine tofu, maple syrup, margarine and vanilla; beat to a creamy consistency.

2. In a separate bowl, combine all-purpose flour, shredded coconut, sugar, almonds, graham crumbs, carob powder, baking powder and salt. Mix well. Gradually stir into tofu mixture until a thick dough is created.

3. With a tablespoon, drop dough onto baking sheet. Bake for 15 minutes. Allow to cool before serving.

PER COOKIE

Calcium	Calories	Protein	Fat	Saturated Fat	Carbohydrates	Fiber	Iron	Sodium
32 mg	109	2 g	4 g	1.6 g	17 g	1 g	1 mg	67 mg

Zucchini Carrot Cake

SERVES 12 TO 14

PREHEAT OVEN TO 325° F (160° C)
11- BY 9-INCH (2.5 L) BAKING DISH

2 cups	grated peeled zucchini	500 mL
1 1/2 cups	granulated sugar	375 mL
1/2 cup	ground almonds	125 mL
3 tbsp	canola oil	45 mL
1/3 cup	maple syrup	75 mL
2 tsp	vanilla	10 mL
1/2 cup	soy milk (calcium fortified)	125 mL
2 cups	cake and pastry flour	500 mL
1 cup	soy flour or amaranth flour	250 mL
1 tsp	baking soda	5 mL
1 tsp	baking powder	5 mL
2 1/2 tsp	cinnamon	12 mL
1/2 tsp	salt	2 mL
2 1/2 cups	grated peeled carrots	625 mL

1. In a large bowl, combine zucchini, sugar, almonds, oil, maple syrup, vanilla and soy milk. Mix well.

2. In a separate bowl, sift together the cake flour, soy flour, baking soda, baking powder, cinnamon and salt. Add to zucchini mixture and, with an electric mixer or by hand, mix until well combined. Fold in carrots.

3. Transfer mixture into prepared baking dish. Bake 55 to 60 minutes, or until a toothpick inserted into the center comes out clean.

4. Cool completely and serve.

Light and tasty, this cake can be enjoyed with or without your favorite frosting. The carrots and zucchini are a nice, healthful calcium addition. It can be served with coffee or tea as a morning or afternoon snack, or as dessert with a side of sweet tofu ice cream (see recipe, page 143).

CALCIUM REPORT

Worldwide, calcium consumption from plant sources is less the exception than the rule.

LAYERED MASHED POTATOES (PAGE 100) ➤
Overleaf: ALMOND PEANUT BUTTER COOKIES (PAGE 124)

				PER SERVING (12)				
Calcium	Calories	Protein	Fat	Saturated Fat	Carbohydrates	Fiber	Iron	Sodium
75 mg	285	7 g	6 g	0.5 g	52 g	2 g	3 mg	234 mg

Orange Kiwi Squares

SERVES 8

PREHEAT OVEN TO 350° F (180° C)
9-INCH (3.5 L) NONSTICK BAKING PAN, WELL GREASED

4 oz	extra firm tofu	125 g
1/2 cup	soy milk (calcium fortified)	125 mL
1/2 cup	chopped peeled orange segments	125 mL
1/2 tsp	vanilla	2 mL
1/2 tsp	almond extract	2 mL
3/4 cup	amaranth flour or soy flour	175 mL
3/4 cup	whole wheat flour	175 mL
1 1/4 cups	brown sugar	300 mL
1 tsp	cinnamon	5 mL
1 tsp	baking soda	5 mL
1/2 tsp	baking powder	2 mL
1/4 tsp	salt	1 mL
3	kiwi fruit, peeled and finely chopped	3
1/4 cup	finely chopped almonds	50 mL
1/2 cup	chocolate chips (dairy-free variety)	125 mL
1/2 cup	carob chips	125 mL

1. In a large bowl with an electric mixer (or in a food processor or blender), combine tofu, soy milk, orange segments, vanilla and almond extract; blend until smooth.

2. In a separate bowl, stir together amaranth flour, whole wheat flour, sugar, cinnamon, baking soda, baking powder and salt. Mix well. Gradually stir dry ingredients into tofu mixture; blend until well mixed. Stir in kiwi, almonds, chocolate and carob chips.

3. Pour batter into prepared baking pan. Bake 1 hour. Serve warm or allow to cool before serving.

◄ HEARTY SOYBEAN BURGERS (PAGE 106)

This dessert combines the flavors of orange, kiwi and chocolate chips. It's a moist and sticky sweet treat that's equally good served warm or at room temperature.

TIP

Great tasting dairy-free chocolate chips can be purchased at health food stores.

PER SERVING								
Calcium	Calories	Protein	Fat	Saturated Fat	Carbohydrates	Fiber	Iron	Sodium
153 mg	348	8 g	8 g	3.5 g	65 g	6 g	3 mg	253 mg

Peanut Butter Marble "Cheesecake"

SERVES 6

PREHEAT OVEN TO 350° F (180° C)
9-INCH (23 CM) NONSTICK PIE PLATE

Hardly anyone can resist the delicious combination of peanut butter, almond butter and chocolate in this creamy "cheesecake." The chocolate is added last and swirled into the tofu to create a marbled appearance that makes this "cheesecake" look and taste like a gourmet delight. The calcium content is also very high — even the pie shell has calcium-rich crushed almonds and maple syrup.

Crust

1 1/2 cups	graham cracker crumbs	375 mL
1/2 cup	maple syrup	125 mL
1/2 cup	ground almonds	125 mL
2 1/2 tbsp	canola oil	30 mL
1/8 tsp	salt	0.5 mL

Filling

8 oz	soft tofu	250 g
8 oz	firm tofu	250 g
1 cup	granulated sugar	250 mL
1/4 cup	maple syrup	50 mL
1/4 cup	orange juice (preferably calcium fortified)	50 mL
1 1/2 tbsp	cornstarch	20 mL
2 tbsp	almond butter	25 mL
2 tbsp	smooth peanut butter	25 mL
1 tbsp	lemon juice	15 mL
2 tsp	vanilla	10 mL

Chocolate Marble

2 1/2 tbsp	cocoa powder	35 mL
2 tsp	maple syrup	10 mL

1. Crust: In a medium bowl, blend graham crumbs, maple syrup, almonds, oil and salt. Press mixture evenly onto greased pie plate and bake 10 minutes. Allow to cool.

PER SERVING								
Calcium	Calories	Protein	Fat	Saturated Fat	Carbohydrates	Fiber	Iron	Sodium
231 mg	602	15 g	23 g	3.2 g	90 g	3 g	6 mg	291 mg

2. In a food processor combine soft and firm tofu, sugar, maple syrup, orange juice, cornstarch, almond butter, peanut butter, lemon juice and vanilla; blend until smooth. Measure out 1/2 cup (125 mL) of the filling and set aside in a small bowl. (This will be used for marbling later.) Pour remaining tofu cream mixture into pie crust.

3. Chocolate marble: In a bowl, combine reserved 1/2 cup (125 mL) tofu filling with cocoa powder and maple syrup; mix well. With a spoon, evenly distribute dollops of chocolate cream (each about 1 tbsp [15 mL]) on the pie filling. Using the tip of a knife, swirl chocolate cream dots into the filling to create a marble effect.

4. Bake 55 minutes. Allow to cool. Refrigerate at least 3 hours before serving.

Figgy Squares

MAKES 12 SQUARES

PREHEAT OVEN TO 350° F (180° C)
9-INCH (2.5 L) NONSTICK BAKING PAN, GREASED

These layered, soft squares can be enjoyed as an after-dinner sweet with tea or coffee. They are also great as a quick calcium snack during the day.

Filling

1 1/2 cups	finely chopped dried figs	375 mL
1/4 cup	raisins	50 mL
1 tbsp	granulated sugar	15 mL
1/2 cup	orange juice (preferably calcium fortified)	125 mL

Crust

1 1/2 tbsp	margarine, softened	20 mL
1 1/4 cups	granulated sugar	300 mL
3/4 cup	all-purpose flour	175 mL
1/2 cup	soy flour	125 mL
1 1/3 cups	soy milk (calcium fortified)	325 mL
1 cup	graham cracker crumbs	250 mL
2 tsp	vanilla	10 mL
1/2 tsp	baking soda	2 mL

TIP

The filling and crust for these squares can be prepared ahead of time. Keep refrigerated until ready to bake.

1. Filling: In a bowl, combine figs, raisins, sugar and orange juice. Cover and refrigerate for 3 hours or overnight.

2. In a food processor or blender, process fig mixture until creamy. Return to bowl and set aside.

3. Crust: In a separate bowl, cream margarine and sugar. In another bowl, sift together all-purpose and soy flours; stir into sugar mixture along with soy milk, graham crumbs, vanilla and baking soda. Mix well.

4. Press half of the crust mixture onto the bottom of the prepared baking pan. Gently spread filling over top. Top off with remaining crust mixture. Bake 35 minutes. Allow to cool before cutting into squares.

		PER SQUARE						
Calcium	Calories	Protein	Fat	Saturated Fat	Carbohydrates	Fiber	Iron	Sodium
96 mg	229	5 g	3 g	0.5 g	48 g	4 g	2 mg	135 mg

Maple Pumpkin Pie

SERVES 6

PREHEAT OVEN TO 350° F (180° C)

12 oz	soft tofu	375 g
1 cup	pumpkin pie filling (canned)	250 mL
1 1/2 tbsp	cornstarch	20 mL
1/4 cup	maple syrup	50 mL
3/4 cup	granulated sugar	175 mL
1 1/2 tsp	vanilla	7 mL
1 tsp	cinnamon	5 mL
1/2 tsp	allspice	2 mL
1	prepared store-bought pie shell (9-inch [23 cm])	1
1/2 cup	pecan halves	125 mL

1. In a food processor, combine tofu, pumpkin, cornstarch, maple syrup, sugar, vanilla, cinnamon and allspice; process until smooth. Pour mixture into prepared pie shell.

2. Decorate top with pecan halves. Bake in preheated oven 45 minutes. Let cool. Chill for 2 to 3 hours before serving.

This pumpkin pie recipe is one of our favorites because it features a delectable combination of creamy tofu and pumpkin pie filling. Even better, it's simple to prepare. You're sure to enjoy this quick and impressive dessert.

TIP

To make tasty pumpkin tarts, simply spoon mixture into pre-made tart shells. This makes a wonderful bite-sized addition to your dessert table. Decorate tarts with a single pecan on each.

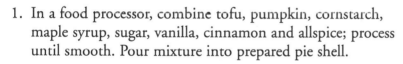

	PER SERVING							
Calcium	Calories	Protein	Fat	Saturated Fat	Carbohydrates	Fiber	Iron	Sodium
178 mg	510	9 g	16 g	2.9 g	91 g	6 g	3 mg	282 mg

Chocolate Chip Cookie Pie

SERVES 6

PREHEAT OVEN TO 325° F (160° C)
9-INCH (23 CM) NONSTICK PIE PLATE, LIGHTLY GREASED

Here's a pie that's guaranteed to satisfy any sweet tooth! With its rich and creamy chocolate chip filling, thick chocolate cookie crust and attractive presentation, you'll be asked to make this tasty, sweet dessert over and over again.

Pie Shell

1 1/4 cups	graham crumbs	300 mL
1/4 cup	ground almonds	50 mL
1/4 cup	shredded coconut	50 mL
1/3 cup	maple syrup	75 mL
1/4 cup	carob powder	50 mL
1/4 cup	whole wheat flour	50 mL
1 tbsp	canola oil	15 mL

Filling

1 lb	firm tofu	500 g
1/2 cup	granulated sugar	125 mL
1/4 cup	maple syrup	50 mL
2 tsp	vanilla	10 mL
3/4 cup	chocolate chips (dairy-free variety)	175 mL

CALCIUM TIPS

Instead of using store-bought shredded coconut, try using the leftover pulp from BASIC ALMOND MILK (see recipe, page 159).

1. In a bowl, combine graham crumbs, almonds, coconut, maple syrup, carob powder, whole wheat flour and oil; mix to form a thick dough.

2. With wet fingers, press dough onto prepared pie plate. (The crust should be thick.) Bake in preheated oven for 15 minutes. Set aside to cool. Increase oven temperature to 350° F (180° C).

3. Make the filling: In a food processor, blend tofu, sugar, maple syrup and vanilla to a fine and creamy texture. Transfer to a bowl. Stir in chocolate chips. Pour mixture into prepared pie shell and bake 35 to 40 minutes. Let cool. Refrigerate 1 hour before serving.

				PER SERVING				
Calcium	Calories	Protein	Fat	Saturated Fat	Carbohydrates	Fiber	Iron	Sodium
218 mg	537	16 g	22 g	7.5 g	78 g	4 g	10 mg	181 mg

Candied Almonds

MAKES 2 1/2 CUPS (625 ML)

2 cups	almonds (unskinned)	500 mL
1/2 cup	water	125 mL
1 cup	granulated sugar	250 mL
4 tsp	vanilla	20 mL

1. In a heavy frying pan over medium heat, stir together all ingredients until bubbling. Continue to cook, stirring occasionally, 10 to 15 minutes or until sugar starts to caramelize around the almonds. As mixture thickens it will begin to form a crust around the almonds. At this point, remove from skillet and transfer almonds to cookie sheet to cool.

This sweet treat is a great source of calcium that can be enjoyed anytime. It's easy to make and requires very few ingredients. In our home, it's one of our favorite snacks — it just disappears too quickly!

TIP

When stirring the almonds it's best to use a plastic or wooden spoon so you don't scrape the bottom of the pan.

PER 1/4 CUP (50 mL)

Calcium	Calories	Protein	Fat	Saturated Fat	Carbohydrates	Fiber	Iron	Sodium
76 mg	250	5.5 g	15 g	1.4 g	26 g	2 g	1 mg	3.5 mg

Sweet Veggie Cupcakes with Almond Butter Frosting

PREHEAT OVEN TO 400° F (200° C)

MAKES ABOUT 20 CUPCAKES
NONSTICK MUFFIN TIN, LIGHTLY OILED

These veggie muffins are rich and tasty — especially with the optional chocolate chips and the nut-butter frosting. Best of all, the disguised beet greens give these muffins a boost of calcium. Even small children will enjoy them.

3 cups	finely chopped beet greens	750 mL
1 1/2 cups	soy milk (calcium fortified)	375 mL
1/4 cup	maple syrup	50 mL
2 tsp	vanilla	10 mL
1 1/2 cups	all-purpose flour	375 mL
1/2 cup	soy flour (preferably defatted)	125 mL
1 1/2 cups	brown sugar	375 mL
1 tbsp	cinnamon	15 mL
1 1/2 tsp	baking soda	7 mL
1 tsp	baking powder	5 mL
1/4 tsp	salt	1 mL
1 cup	chocolate chips (optional)	250 mL

1. Steam beet greens. Allow to cool. Transfer to a large bowl along with soy milk, maple syrup and vanilla; mix together.

2. In a separate bowl, sift together flours, brown sugar, cinnamon, baking soda, baking powder and salt. Add in the chocolate chips, if using. Gradually stir dry ingredients into beet-green mixture; mix well.

3. Spoon mixture into cups of prepared muffin tin. Bake 20 minutes, checking for doneness with a toothpick (insert toothpick into center; if it comes out clean, it is done). Remove from oven and let cool. Top with ALMOND BUTTER FROSTING. (See next page.)

PER CUPCAKE								
Calcium	Calories	Protein	Fat	Saturated Fat	Carbohydrates	Fiber	Iron	Sodium
80 mg	142	3 g	3 g	0.3 g	28 g	1 g	2 mg	152 mg

Almond Butter Frosting

MAKES 1 CUP (250 mL)

4 oz	regular tofu **or** firm tofu	125 g
1/2 cup	brown sugar	125 mL
1/4 cup	almond butter, softened	50 mL
2 tbsp	maple syrup	25 mL
1/2 tbsp	peanut butter	7 mL
1/2 tsp	vanilla	2 mL

1. In a food processor or blender, combine all ingredients; blend until creamy. Transfer to a small bowl and refrigerate for 1 hour. When frosting has become firm, it's ready to spread.

				PER 2 TBSP (25 mL)				
Calcium	Calories	Protein	Fat	Saturated Fat	Carbohydrates	Fiber	Iron	Sodium
51 mg	130	3 g	6 g	0.6 g	19 g	1 g	1 mg	11 mg

Top any of your favorite muffins, cakes, cupcakes or fruit with this creamy, sweet frosting. Kids love the rich nutty flavor.

Sweet Sesame Almond Crumble

MAKES 1 CUP (250 mL)

3/4 cup	almonds, unskinned	175 mL
2	brazil nuts	2
2 tbsp	brown sugar	25 mL
2 tbsp	sesame seeds	25 mL
1 tsp	cinnamon	5 mL
1/8 tsp	nutmeg	0.5 mL

1. In a small bowl, stir together all ingredients. Transfer to a blender or food processor; blend to a fine, crumbly consistency.

				PER 2 TBSP (25 mL)				
Calcium	Calories	Protein	Fat	Saturated Fat	Carbohydrates	Fiber	Iron	Sodium
40 mg	102	3 g	8 g	0.9 g	6 g	1 g	1 mg	3 mg

Sprinkle this sweet, crunchy topping over cereal, fruit, pudding, or your favorite frozen dessert. It even goes well with baked potatoes!

TIP

To maintain freshness, store nuts in refrigerator.

Almond Mocha Cream Tarts

Makes 48 mini tarts

Spoon this delicious creamy pudding into pre-made mini tart shells or pour into fancy dessert cups. If serving to children, you can leave out the coffee grains. Or eliminate the caffeine by using a coffee substitute (available at your local health food store).

1/2 cup	blanched almonds	125 mL
1/2 cup	soy milk (calcium fortified)	125 mL
8 oz	firm tofu	250 g
4 oz	soft tofu	125 g
3/4 cup	granulated sugar	175 mL
1 tbsp	cocoa powder	15 mL
2 tbsp	chocolate chips (dairy-free variety)	25 mL
1/2 tbsp	instant coffee granules (or coffee substitute)	5 mL
1/2 tbsp	blackstrap molasses	5 mL
1 1/2 tsp	vanilla	7 mL
1	package pre-made mini tart shells	1

TIPS

Increase the amount of coffee (or substitute) to get the full coffee flavor you prefer.

To blanch almonds, soak almonds in a bowl of boiling water for 2 minutes. Peel off and discard skins.

1. In a blender or food processor, purée almonds and soy milk until creamy and smooth. Pour mixture through a strainer into a bowl, pressing the solids with the back of a spoon. Discard solids.

2. Rinse out blender and add strained "almond milk." Add tofu and sugar; blend until creamy and smooth.

3. Transfer mixture to large saucepan along with cocoa powder, chocolate chips, coffee granules, molasses and vanilla. Whisk together over low heat for 2 minutes.

4. Spoon almond mocha cream into mini tart shells. Refrigerate for 1 1/2 hours before serving.

	PER MINI TART							
Calcium	Calories	Protein	Fat	Saturated Fat	Carbohydrates	Fiber	Iron	Sodium
25 mg	99	2 g	6 g	1.3 g	10 g	0 g	1 mg	82 mg

Dreamy Chocolate Mousse

SERVES 4

MAKES 2 CUPS (500 mL)

8 oz	*soft tofu*	*250 g*
4 oz	*regular tofu* or *firm tofu*	*125 g*
3 tbsp	*granulated sugar*	*45 mL*
1/2 tsp	*vanilla*	*2 mL*
1 cup	*chocolate chips (dairy-free variety)*	*250 mL*

1. In a blender or food processor, blend tofu, sugar and vanilla until smooth. Set aside.

2. In a small saucepan over low heat, slowly melt chocolate chips. Transfer immediately to mixture in blender. Blend until creamy and smooth.

3. Pour mousse into small dessert cups or bowls. Refrigerate 1 to 2 hours or until mousse sets. Serve cold.

Indulge in this rich and luscious mousse. A wonderfully versatile dessert, use it to fill a pie or mini tarts. Once refrigerated, it will become firm and easy to slice. Your family and guests are guaranteed to enjoy this calcium-enhanced, chocolaty dessert.

TIPS

If using mousse as a pie or tart filling, you may need to bake the shell first. Check package directions.

You can also use the mousse as a topping for cupcakes or as a cake icing — simply let the mousse set in the refrigerator before using.

		PER SERVING						
Calcium	Calories	Protein	Fat	Saturated Fat	Carbohydrates	Fiber	Iron	Sodium
136 mg	312	8 g	18 g	9 g	36 g	3 g	3 mg	17 mg

Butterscotch Pudding

SERVES 4

8 oz	*soft tofu*	250 g
6 oz	*firm tofu*	175 g
1 1/2 cups	*butterscotch chips (dairy-free variety)*	375 mL

1. In blender or food processor, blend soft and firm tofu to a creamy consistency.

2. In a small saucepan over low heat, slowly melt butterscotch chips. Transfer immediately to mixture in blender. Blend until creamy and smooth.

3. Pour pudding into small dessert cups or bowls. Refrigerate 1 to 2 hours or until pudding sets. Serve cold.

Imagine a wonderfully creamy butterscotch pudding in 3 simple steps. A hit with children and adults alike, serve it alone in small decorative bowls or spoon into pre-made chocolate cups (usually found in specialty food shops); refrigerate for a few hours and you'll have yourself a gourmet dessert!

TIPS

For a festive appearance, decorate pudding with chocolate shavings or colored sprinkles.

This pudding also makes an incredible butterscotch ice cream — just freeze at least 2 hours.

PER SERVING

Calcium	Calories	Protein	Fat	Saturated Fat	Carbohydrates	Fiber	Iron	Sodium
139 mg	401	11 g	22 g	16.0 g	41 g	0.3 g	2 mg	18 mg

Banana Pudding

SERVES 4

MAKES 3 CUPS (750 mL)

10 oz	soft tofu	300 g
6 oz	firm tofu	175 g
2	small very ripe bananas	2
1/2 cup	granulated sugar	125 mL
1/4 cup	soy milk (calcium fortified)	50 mL
2 tsp	vanilla	10 mL

1. In a blender or food processor, blend all ingredients until creamy and smooth. Pour into small dessert cups and refrigerate for 2 hours. Serve cold.

If you love bananas, then you will appreciate this smooth and silky sweet banana pudding. Kids of all ages also enjoy this tasty calcium-rich dessert — whether on its own or as a topping over fruit.

TIP

Garnish pudding with a slice of fresh banana on top or thin slices around the edges of the bowl.

	PER SERVING							
Calcium	Calories	Protein	Fat	Saturated Fat	Carbohydrates	Fiber	Iron	Sodium
226 mg	260	12 g	6 g	0.9 g	41 g	1 g	5 mg	20 mg

Strawberry "Yogurt" Delight

SERVES 4

MAKES 2 1/2 CUPS (625 mL)

2 cups	whole frozen strawberries	500 mL
8 oz	firm tofu	250 g
1/3 cup	granulated sugar	75 mL
	Fresh mint leaves	
	Fresh strawberries	

1. In a blender or food processor, blend strawberries, tofu and sugar until smooth. Transfer to dessert bowls or parfait glasses. Refrigerate 1 hour. Serve cold, garnished with a few mint leaves and a slice of strawberry on top.

Creamy Strawberry "Yogurt" Popsicles

1. Prepare recipe above and spoon yogurt into plastic popsicle molds. Freeze at least 4 hours before serving. To remove popsicle from the container, run hot water over the plastic mold for 30 seconds.

Try this light, semi-sweet strawberry "yogurt" alone or spooned over your favorite mixed fruit or pie. We developed this recipe because our daughter loves the flavor and consistency of fruity yogurt. It's simple to make and more than a few times we've served it as a main dish for our daughter on her picky "I don't want to eat" days.

CALCIUM TIP

Although soft tofu works well in creamy dessert recipes, firm tofu was chosen here because it's higher in calcium.

	Calcium	Calories	Protein	Fat	Saturated Fat	Carbohydrates	Fiber	Iron	Sodium
PER SERVING	128 mg	157	9 g	5 g	0.7 g	22 g	2 g	7 mg	10 mg

Sweet Chocolate "Ice Cream"

SERVES 2

MAKES 2 CUPS (500 ML)

1	very ripe banana	1
6 oz	soft tofu	180 g
1/2 cup	soy milk (calcium fortified)	125 mL
1 tbsp	granulated sugar	15 mL
1/4 cup	maple syrup	50 mL
1/4 cup	raisins	50 mL
3 tsp	cocoa	15 mL

1. In a blender or food processor, blend all ingredients until smooth. Transfer to a plastic bowl. Freeze at least 3 hours.

2. Remove from freezer and place in food processor. Blend until creamy. Serve immediately.

This frozen dessert is a delicious cold treat. You'll be surprised at how closely it resembles real ice cream! It can be lots of fun for kids if you make a face on top with raisins or chopped figs.

CALCIUM NEWS

In Western countries, high salt and protein consumption contribute to calcium loss. To compensate, many government food guides recommend daily calcium intakes of 1000 mg or more. Yet, the World Health Organization (WHO) recommends only 400 to 500 mg daily.

				PER SERVING				
Calcium	Calories	Protein	Fat	Saturated Fat	Carbohydrates	Fiber	Iron	Sodium
254 mg	310	8 g	4 g	0.6 g	64 g	3 g	1 mg	28mg

Cream Whip in Papayas with Kiwi and Oranges

SERVES 6

Papayas filled with a tasty, mint-flavored cream whip, topped with fresh fruit — all combine to make this a wonderfully refreshing dessert. Kiwi and oranges provide an added calcium punch. These fruits are also appealing in both texture and color — especially with the bright orange papayas.

8 oz	regular or soft tofu	250 g
Half	small ripe banana	Half
1/4 cup	granulated sugar	50 mL
2 tbsp	maple syrup	25 mL
1 tbsp	almond butter	15 mL
1/4 tsp	mint extract	1 mL
3	large papayas	3
4	kiwis, peeled and chopped	4
1	orange, peeled and chopped	1

1. In a blender or food processor, combine tofu, banana, sugar, maple syrup, almond butter and mint extract; blend until smooth and thick. Refrigerate for 1 hour or up to 2 days.

2. With a sharp knife, cut papayas in half; remove and discard seeds. Fill papayas with cream mixture. Garnish with kiwi and oranges. Serve immediately.

TIP

For a zesty crunch and more calcium, sprinkle this dish with some SWEET SESAME ALMOND CRUMBLE (see recipe, page 137).

	PER SERVING							
Calcium	Calories	Protein	Fat	Saturated Fat	Carbohydrates	Fiber	Iron	Sodium
111 mg	203	5 g	4 g	0.5 g	41 g	5 g	3 mg	11 mg

Nutty Soybean Treat

SERVES 4

1 cup	dried soybeans	250 mL
1/8 tsp	canola oil	0.5 mL
1/8 tsp	garlic salt	0.5 mL

1. Using a strainer under cold running water, rinse soybeans. Transfer to a large bowl, cover with water and soak 8 hours or overnight. Drain and pat dry.

2. In a nonstick skillet, cook soybeans over high heat, shaking the skillet back and forth, as if making popcorn. Continue for approximately 3 to 4 minutes, then reduce heat to medium-high; cook, still shaking the skillet periodically, for another 8 minutes or until the beans have browned.

3. In a small bowl, stir together oil and garlic salt. Add to skillet and cook, still shaking, another 2 minutes. Allow beans to cool 5 minutes before serving.

Soybeans are not only high in calcium, but they're also incredibly versatile. Cooked, they're a great addition to soups, salads and other dishes. Pan roasted, as here, they turn into a fun, nutty snack. We have found that children love this healthy treat just as much as adults!

PER SERVING

Calcium	Calories	Protein	Fat	Saturated Fat	Carbohydrates	Fiber	Iron	Sodium
114 mg	195	19 g	10 g	1.5 g	11 g	6 g	6 mg	73 mg

Cocoa Pancakes

Makes 10 to 12 pancakes

2 cups	soy milk (calcium fortified)	500 mL
1/3 cup	granulated sugar	75 mL
1 1/2 tbsp	blackstrap molasses	20 mL
1 cup	all-purpose flour	250 mL
3/4 cup	amaranth flour or soy flour	175 mL
2 1/2 tbsp	carob powder	35 mL
1 tsp	baking powder	5 mL
1/2 tsp	baking soda	2 mL

1. In a large bowl, combine soy milk, sugar and molasses. Mix well.

2. In a separate bowl, sift together all-purpose flour, amaranth flour, carob powder, baking powder and baking soda. Gradually stir into soy milk mixture until a thick batter forms.

3. Heat a lightly oiled nonstick skillet over medium heat. Drop approximately 2 tbsp (25 mL) batter onto hot skillet for each pancake. For a thinner pancake, spread batter with the back of a spoon. Cook 2 minutes on each side.

Everyone loves these semi-sweet pancakes. They contain just enough carob flour and molasses to create exciting and tasty pancakes — something different that can be served anytime. Our young daughter and her friends love to eat them as a snack.

	PER PANCAKE (10)							
Calcium	Calories	Protein	Fat	Saturated Fat	Carbohydrates	Fiber	Iron	Sodium
119 mg	131	4 g	2 g	0.3 g	26 g	2 g	2 mg	96 mg

Rich and Simple Pancakes

MAKES 20 TO 24 PANCAKES

1 1/4 cups	water	300 mL
8 oz	soft tofu	250 g
1/4 cup	maple syrup	50 mL
2 tbsp	canola oil	25 mL
1/4 cup	finely chopped dried figs	50 mL
1 1/2 cups	whole wheat flour	375 mL
1/2 cup	amaranth flour or soy flour	125 mL
1 tsp	baking powder	5 mL
1 tsp	ground allspice	5 mL
1/2 tsp	cinnamon	2 mL

1. In a blender or food processor, blend water, tofu, maple syrup and oil until milky. Set aside.

2. In a small bowl, cover figs with 1 tbsp (15 mL) boiling water and let sit 10 minutes. Add to tofu mixture in blender; blend until well combined. Transfer to a large bowl.

3. In a separate bowl, sift together the whole wheat flour, amaranth flour, baking powder, allspice and cinnamon. Add to tofu mixture and combine until blended.

4. Heat a lightly oiled nonstick skillet over medium heat. Spoon approximately 1 1/2 tbsp (20 mL) of batter onto hot skillet. Spread batter out to form pancake. Cook on one side for about 2 minutes. Check for doneness (edges should be golden brown) then flip over and cook about 1 1/2 minutes more. Serve with your favorite topping or maple syrup.

These hearty pancakes look like the traditional variety, but they are made with calcium-rich amaranth flour, tofu and figs. Kids love them because they have a delicious cinnamon-sweet flavor — even before the maple syrup is poured on top!

CALCIUM FACT

People who consume less protein and salt tend to retain more calcium.

				PER PANCAKE (20)				
Calcium	Calories	Protein	Fat	Saturated Fat	Carbohydrates	Fiber	Iron	Sodium
42 mg	84	3 g	2 g	0.2 g	15 g	2 g	1 mg	17 mg

Beverages

Beverages are deliciously versatile foods, whether as accompaniments to other dishes or as calcium-rich between-meal snacks. Here you'll find recipes for a variety of beverages — from rich, thick (and filling) smoothies to light refreshing juices.

These recipes are also great for children.

You can make the milks or smoothies more festive by shaving chocolate or sprinkling cinnamon on top. Try adding a sliced strawberry or orange propped on the side of the glass to make the drink sparkle with color. Or serve the juices in a crystal goblet and enjoy your calcium boost with style!

SMOOTHIES

Smoothies are a delicious and nutritious way to encourage children
(and adults) to consume more fruits and calcium-rich ingredients. These include
calcium-fortified soy milk and orange juice, molasses, carob, fruits and nuts — all
used here to create healthy, calcium-rich alternatives to traditional shakes.
Because bananas are used so often to give volume to smoothies,
it makes sense to keep some frozen bananas in the freezer.
(See below for tips on preparing and freezing bananas.)
Strawberries, blueberries or other fruit are also useful to keep in
the freezer as flavor enhancers.

Basic Banana Smoothie

MAKES 3 CUPS (750 mL)

This basic banana smoothie delivers its calcium from the fortified soy milk, tofu and blackstrap molasses. The molasses also gives this drink a delicious malt flavor. We frequently have this in place of our morning cereal.

2	frozen ripe bananas*	2
1 1/2 cups	soy milk (calcium fortified)	375 mL
4 oz	soft tofu	125 g
1 tbsp	blackstrap molasses	15 mL
2 tbsp	maple syrup	25 mL

1. Cut frozen bananas into 1-inch (2.5 cm) pieces. Transfer to
   blender along with soy milk, tofu, molasses and maple
   syrup; blend until smooth. Enjoy immediately or refrigerate
   until ready to serve.

* *Freezing bananas*: Allow bananas to ripen so that dark spots
   appear on the peel. Peel bananas and place them whole in a
   plastic container or plastic bag. Seal container or bag and
   freeze at least 5 hours.

PER CUP (250 mL)								
Calcium	Calories	Protein	Fat	Saturated Fat	Carbohydrates	Fiber	Iron	Sodium
283 mg	181	6 g	4 g	0.5 g	33 g	3 g	2 mg	29 mg

"Chocolate" Malted Banana Smoothie

MAKES 2 CUPS (500 ML)

2	*frozen ripe bananas (see opposite page)*	2
1 cup	*cold soy milk (calcium fortified)*	250 mL
1/2 tbsp	*blackstrap molasses*	7 mL
1/2 tbsp	*carob powder*	7 mL

*I*n this high-energy, calcium-rich drink, carob powder replaces cocoa to create a thick, rich and tasty chocolate flavor.

1. Cut frozen bananas into 1-inch (2.5 cm) pieces. Add to blender along with soy milk, molasses and carob powder; blend until thick and smooth.

2. Wrap 2 to 4 ice cubes in a clean towel and break them up slightly with a heavy pan. Add ice chips to blender and blend until smooth. Enjoy immediately or refrigerate until ready to serve. Mix again with a spoon before serving.

TIP

For a sweet and decorative touch, sprinkle smoothie with icing sugar.

	PER CUP (250 ML)							
Calcium	Calories	Protein	Fat	Saturated Fat	Carbohydrates	Fiber	Iron	Sodium
207 mg	159	5 g	3 g	0.5 g	33 g	4 g	2 mg	21 mg

Orange Banana Dream Smoothie

MAKES 3 CUPS (750 ML)

This semi-sweet orange and banana drink is even more delicious when chilled. Oranges provide a slight increase in calcium, as well as a frothy consistency. Increase sweetness and calcium with additional maple syrup.

2	*small frozen ripe bananas (see page 150)*	2
2	*seedless oranges* or *3 mandarin oranges, peeled and cut into pieces*	2
1 1/4 cups	*soy milk (calcium fortified)*	300 mL
3 tbsp	*maple syrup*	45 mL

1. Cut frozen bananas into 1-inch (2.5 cm) pieces and place in blender along with orange pieces, soy milk and maple syrup; blend until smooth and frothy. Enjoy immediately or refrigerate until ready to serve.

*T*IP

For a thicker, colder consistency when blended, freeze the pieces of orange in advance. Add a slice of orange to the side of the glass to create a more festive appearance.

				PER CUP (250 mL)				
Calcium	Calories	Protein	Fat	Saturated Fat	Carbohydrates	Fiber	Iron	Sodium
176 mg	181	4 g	2 g	0.3 g	39 g	4 g	1 mg	15 mg

Fruity Smoothie

MAKES 2 1/2 CUPS (625 mL)

1	large seedless orange, peeled and chopped	1
1 cup	fruit juice (preferably calcium-fortified, such as orange juice)	250 mL
1/2 cup	halved strawberries (fresh or frozen)	125 mL
4 oz	soft tofu	125 g
3 tbsp	maple syrup	45 mL
1 1/2 tsp	vanilla	7 mL

1. In a blender, combine all ingredients; blend on high for 30 seconds. Refrigerate until cold or add crushed ice and serve immediately.

This refreshing fruity drink can be sweetened further by increasing the maple syrup, which will also provide an increase in calcium. Bump up the volume of this drink a little by adding crushed ice.

TIP

For an attractive presentation, place a single strawberry on the side of the glass. Slice strawberry part way down the middle and place on the rim.

PER 1 1/4 CUPS (300 mL)

Calcium	Calories	Protein	Fat	Saturated Fat	Carbohydrates	Fiber	Iron	Sodium
296 mg	219	5 g	2 g	0.3 g	45 g	3 g	1 mg	14 mg

MILKS

Nut milks are one of the best high-calcium, nutrient-packed alternatives to dairy milk, and can be used for everything from cereals to soups. Experiment with these rich milks in your cooking — particularly those based on almonds or sesame seeds. All you need to make nut milk is a handful of nuts or seeds, some water and a blender.

To add more interest and to boost the calcium further, try adding molasses and carob. Keep in mind that maple syrup not only adds sweetness but also adds calcium to homemade nut milks.

If you're looking for convenience and you want to get the highest calcium content per recipe, be sure to purchase calcium-fortified non-dairy milks. See Calcium Table (page 174). By drinking a couple of these each day, you're easily on your way to getting your daily calcium requirement.

This is one of the simplest and fastest ways to create "milk." If you wish, add a sweetener of your choice — either maple syrup or brown sugar.

TIP

For a delicious flavor variation, simply replace vanilla with 1/4 tsp (1 mL) mint, butter-scotch or almond extract.

Basic Sesame Milk

MAKES 2 CUPS (500 ML)

1/4 cup	sesame seeds	50 mL
2 cups	cold soy milk (calcium fortified)	500 mL
3 tbsp	maple syrup	45 mL
1/2 tsp	vanilla	2 mL

1. Using a fine strainer, rinse sesame seeds under cold water; drain. Transfer to blender along with soy milk; blend until smooth.

2. Pour milk through a strainer into a small pitcher. Press solids with the back of a spoon. Discard solids.

3. Rinse blender and add strained milk. Add maple syrup and vanilla; blend a few seconds until well mixed. Refrigerate and serve cold.

				PER CUP (250 mL)				
Calcium	Calories	Protein	Fat	Saturated Fat	Carbohydrates	Fiber	Iron	Sodium
286 mg	214	9 g	12 g	1.5 g	21 g	3 g	3 mg	32 mg

Rich Sesame Milk

MAKES 2 CUPS (500 mL)

1/4 cup	sesame seeds	50 mL
1 1/2 cups	cold soy milk (calcium fortified)	375 mL
Half	ripe banana	Half
2 1/2 tbsp	maple syrup	35 mL

1. Using a fine strainer, rinse sesame seeds under cold water; drain. Add to blender along with soy milk; blend on high until smooth.

2. Pour milk through a strainer into a small pitcher. Press solids with the back of a spoon. Discard solids.

3. Rinse blender and add strained milk along with banana and maple syrup. Blend until smooth. Refrigerate and serve cold.

The soy milk, banana and sesame seeds in this mixture make a deliciously rich drink. It's a great calcium-enhanced milk to pour over cereal or to use in baking.

TIP

For an interesting, flavorful garnish, try a sprinkle of nutmeg.

PER CUP (250 mL)

Calcium	Calories	Protein	Fat	Saturated Fat	Carbohydrates	Fiber	Iron	Sodium
224 mg	207	8 g	11 g	1.5 g	23 g	3 g	2 mg	26 mg

"Chocolate" Milk

Deliciously sweet, this calcium-rich drink features a healthful blend of carob and almonds. It's a great alternative to conventional chocolate milk.

1/4 cup	*blanched almonds (see Tip, page 138, for technique)*	*50 mL*
2 cups	*soy milk (calcium fortified)*	*500 mL*
1/4 cup	*maple syrup*	*50 mL*
1 tbsp	*carob powder*	*15 mL*
1/2 tsp	*vanilla*	*2 mL*

1. In a blender, combine almonds and soy milk; blend to a fine liquid. Pour the milk through a fine strainer into a small pitcher. Press solids with the back of a spoon. Discard solids.

2. Rinse blender and add strained milk along with maple syrup, carob powder and vanilla; blend until well mixed. Refrigerate and serve cold.

CALCIUM TIP

Check your local grocer or health food store for calcium-fortified milk alternatives. Test a variety of fortified grain milks — any one will work wonderfully in the recipes that require soy milk. See Calcium Table (page 174) for calcium-fortified milk alternatives.

	PER CUP (250 mL)							
Calcium	Calories	Protein	Fat	Saturated Fat	Carbohydrates	Fiber	Iron	Sodium
310 mg	228	8 g	10 g	1.1 g	30 g	3 g	2 mg	29 mg

Frothy Hot Chocolate

MAKES 1 CUP (250 mL)

1 cup	soy milk (calcium-fortified)	250 mL
2 tsp	carob powder or cocoa	10 mL
1 1/2 tsp	maple syrup	7 mL
1/4 tsp	vanilla or mint extract (optional)	1 mL

1. In a small saucepan, heat soy milk over medium heat until a couple of bubbles appear. Remove from heat. Stir in carob powder, maple syrup and vanilla. Beat with wire whisk or transfer to blender and blend until frothy.

Calcium-fortified milk alternatives are delicious served hot with added flavorings. One of our favorites is hot chocolate with mint extract. Blend long enough and you'll create a thick foam that will settle on top. Make yourself an incredible hot mocha by blending in a little coffee — it's a gourmet delight!

TIP

Make this drink into a coffee house specialty by adding dark chocolate shavings over top.

PER CUP (250 mL)

Calcium	Calories	Protein	Fat	Saturated Fat	Carbohydrates	Fiber	Iron	Sodium
319 mg	110	7 g	5 g	0.5 g	14 g	3g	2 mg	31 mg

Flavored Cappuccino

MAKES 1 CUP (250 ML)

1 cup	soy milk (calcium fortified)	250 mL
1 tsp	coffee (or coffee substitute)	5 mL
1 1/2 tsp	granulated sugar	7 mL
1/4 tsp	mint, butterscotch or almond extract	2 mL

1. In a small saucepan, heat soy milk over medium heat until a couple of bubbles appear. Remove from heat. Stir in coffee, sugar and mint or other flavoring. Beat with wire whisk or transfer to blender and blend until frothy.

There is no better way to enjoy coffee than with a bit of added flavoring. The warmed milk is sweetly fragrant and it foams up beautifully in the blender. It will remind you of that cozy café feeling on those cold or rainy days.

TIP

Try this recipe with and without the flavorings to see which way you prefer it.

PER CUP (250 mL)

Calcium	Calories	Protein	Fat	Saturated Fat	Carbohydrates	Fiber	Iron	Sodium
301 mg	112	7 g	5 g	0.5 g	12 g	3 g	1 mg	29 mg

Basic Almond Milk

MAKES 2 CUPS (500 ML)

1/2 cup	*blanched almonds (see Tip, page 138, for technique)*	*125 mL*
2 cups	*cold water*	*500 mL*
2 tbsp	*maple syrup*	*25 mL*

1. In a blender combine almonds and water; blend to a fine liquid. Pour milk through a fine strainer into a small pitcher. Press solids with the back of a spoon. Discard solids. Rinse blender and add strained milk along with maple syrup; blend until well mixed. Refrigerate and serve cold.

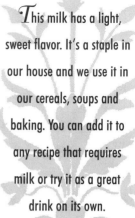

This milk has a light, sweet flavor. It's a staple in our house and we use it in our cereals, soups and baking. You can add it to any recipe that requires milk or try it as a great drink on its own.

CALCIUM TIP

The leftover pulp from any of the milk recipes can be used as a calcium-boosting ingredient in muffins or burgers.

PER CUP (250 mL)

Calcium	Calories	Protein	Fat	Saturated Fat	Carbohydrates	Fiber	Iron	Sodium
76 mg	186	5 g	13 g	1.2 g	15 g	2 g	1 mg	10 mg

Rich Almond Coconut Milk

Makes 1 1/3 cups (325 mL)

1/4 cup	blanched almonds (store-bought or see Tip, page 138, for technique)	50 mL
1 cup	soy milk (calcium fortified)	250 mL
1/2 cup	light coconut milk (canned)	125 mL
3 tbsp	maple syrup	45 mL

1. In a blender, combine almonds, soy milk and coconut milk; blend to a fine liquid. Pour mixture through a fine strainer into a small pitcher. Press solids with the back of a spoon. Discard solids.

2. Rinse blender and add strained milk along with maple syrup; blend until well mixed. Refrigerate and serve cold.

This beverage is rich, but has a light, sweet flavor. Try it with cereals or drink it on its own. Either way, it will boost your daily calcium intake.

CALCIUM QUOTE

"It's quite clear that there's better sources of calcium [than cow's milk]. We don't need cow's milk to get our calcium."

Frank A. Oski, M.D.

Interview - Jan.11, 1996

Peanut Butter Marble "Cheesecake" (Page 130) ➤

PER CUP (250 mL)								
Calcium	Calories	Protein	Fat	Saturated Fat	Carbohydrates	Fiber	Iron	Sodium
229 mg	273	8 g	16 g	4.1 g	29 g	3 g	2 mg	42 mg

JUICES

With any kind of electric juice extractor, it is simple to create delicious calcium-rich drinks from a wide variety of fruit or vegetables. While some of these fruits and vegetables do not contain much calcium on their own, the amount of calcium increases significantly when concentrated or combined with other ingredients.

Drink freshly made juice as soon as possible after making it. Juice kept too long in the fridge will lose valuable nutrients. It also tastes best when fresh.

Taste-of-Grape Juice

MAKES 1 CUP (250 ML)

2	small sweet apples, cored and thickly sliced	2
1 cup	beet greens (about 8 to 10 leaves)	250 mL
Half	beet (root), scrubbed and quartered	Half

1. Add a few slices of apple to juicer, along with 2 or 3 beet greens and one quarter of the beet; extract juice. Repeat this process until ingredients are used up. (Finish off with apple slices to make the juice flow smoothly through the juicer.) Discard solids.

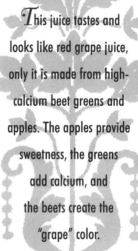

This juice tastes and looks like red grape juice, only it is made from high-calcium beet greens and apples. The apples provide sweetness, the greens add calcium, and the beets create the "grape" color.

TIP

For best juice results use Red Delicious apples. They create a tastier, sweeter juice.

◄ STRAWBERRY "YOGURT" DELIGHT (PAGE 142)

				PER CUP (250 mL)				
Calcium	Calories	Protein	Fat	Saturated Fat	Carbohydrates	Fiber	Iron	Sodium
94 mg	115	2 g	1 g	0.1 g	29 g	7 g	3 mg	162 mg

Fresh Collard-Apple Juice

MAKES 1 CUP (250 mL)

2	sweet apples, cored and quartered	2
1 1/4 cups	packed collard greens (leaves and stems), rinsed	300 mL

1. Add a few slices of apple to juicer then some of the collard greens. Extract juice. Repeat this process until ingredients are used up. (Finish off with apple slices to make juice flow smoothly through the juicer.) Discard solids.

The apples in this calcium-enhanced juice provide a tangy yet sweet taste. Try different varieties of apples for a slight change of flavor.

CALCIUM TIP

A glass of fortified orange juice contains as much or more calcium as the same quantity of milk – but without the fat and animal protein.

PER CUP (250 mL)

Calcium	Calories	Protein	Fat	Saturated Fat	Carbohydrates	Fiber	Iron	Sodium
72 mg	131	1 g	1 g	0.1 g	34 g	5 g	0 mg	9 mg

Carrot Juice

MAKES 1 1/2 CUPS (375 mL)

8 to 10	medium carrots, well scrubbed or peeled	8 to 10
1 tbsp	maple syrup (optional)	15 mL

1. Add carrots to juicer. Extract juice and discard solids. Empty juice into a glass, add maple syrup and stir until well mixed. Refrigerate or serve immediately.

So simple to make, this basic carrot juice tastes best if you use sweet carrots.

	PER CUP (250 mL)							
Calcium	Calories	Protein	Fat	Saturated Fat	Carbohydrates	Fiber	Iron	Sodium
66 mg	105	3 g	0.5 g	0.1 g	23 g	2 g	1 mg	86 mg

Carrot Kale Apple Juice

MAKES 1 1/2 CUPS (375 ML)

1 1/2 cups	packed kale, rinsed	375 mL
5	sweet carrots, peeled	5
1	Red Delicious apple, cored and quartered	1

1. Add some of each of the kale, carrots and apple to the juicer. Extract juice. Repeat this process until ingredients are used up. Discard solids. Refrigerate or add ice and serve immediately.

The carrot and apple in this high-calcium juice offset the slightly bitter flavor of kale.

PER CUP (250 mL)

Calcium	Calories	Protein	Fat	Saturated Fat	Carbohydrates	Fiber	Iron	Sodium
78 mg	118	3 g	1 g	0.1 g	27 g	3 g	1 mg	64 mg

Lemon Orange Juice

MAKES 1 1/2 CUPS (375 mL)

1	seedless orange, peeled (but with some of the white pith remaining) and quartered	1
1 cup	orange juice (preferably calcium fortified)	250 mL
Half	fresh lemon, juice only	Half

1. In a blender, liquefy orange pieces. Add orange juice and lemon juice; blend until well mixed. Strain the juice and discard pulp. Refrigerate or add ice and enjoy immediately.

You can enjoy a glass of calcium-fortified orange juice on its own for a quick calcium boost. Or try this recipe with ice for a deliciously cold, refreshing drink.

	PER CUP (250 mL)							
Calcium	Calories	Protein	Fat	Saturated Fat	Carbohydrates	Fiber	Iron	Sodium
228 mg	106	2 g	1 g	0.1 g	25 g	2 g	0 mg	2 mg

Cranberry Papaya Juice

MAKES 2 CUPS (500 mL)

1 cup	orange juice	250 mL
	(preferably calcium fortified)	
Half	ripe papaya, seeded, peeled and chopped	Half
1/2 cup	cranberry juice	125 mL
1 1/2 tsp	lemon juice	7 ml

1. In a blender, process all ingredients until smooth. Refrigerate and serve cold.

PER CUP (250 mL)

Calcium	Calories	Protein	Fat	Saturated Fat	Carbohydrates	Fiber	Iron	Sodium
167 mg	115	1 g	0 g	0.1 g	28 g	1 g	0 mg	6 mg

Appendix A
Calcium Boosters

Almond Butter Bean Spread

There is no better way to disguise beans than with the delicious flavor of almond butter. Mash 1/3 cup (75 mL) cooked navy beans. Mix together with 1/4 cup (50 mL) almond butter, 1 tbsp (15 mL) water and 1 tsp (5 mL) maple syrup. (You can make more or less by simply adjusting the quantities proportionally.)

Almond Butter Tahini Spread

Tahini alone can be bland. With the addition of almond butter, however, it becomes very appealing. Try blending 2 tbsp (25 mL) of tahini with 1/4 cup (50 mL) almond butter along with 1 tbsp (15 mL) of water. Mix in 1 tsp (5 mL) sugar for some sweetness. (Note: the water helps to achieve a softer texture; spread over crackers or scoop into mini pita bread.)

Crunchy Peanut Fig Butter

To create a crunchy texture, chop 6 fresh figs into tiny pieces and add to 1/2 cup (125 mL) peanut or almond butter. Mix together well.

Peanut Butter Tofu Spread

Tofu is one ingredient which absorbs the flavoring of anything it is mixed with so it will be difficult to notice any difference when blended with the strong flavor of peanut or almond butter. Mash 3 tbsp (45 mL) peanut or almond butter to 2 tbsp (25 mL) firm tofu and 1 tsp (5 mL) maple syrup.

Cereal

Add ground almonds, brazil nuts, sesame seeds, raisins or chopped dried figs to any hot or cold cereal — or cooked amaranth grains (see page 20 for technique) — with maple syrup to taste. You might even try a combination of all of these together with half the amount of cereal to fill an entire bowl. Use a calcium-fortified milk alternative whenever you can.

SALAD

Salads can be more interesting and calcium-enriched by adding finely chopped and lightly steamed vegetables such as broccoli, zucchini, and especially shredded greens such as collards, beet greens or kale. Cool the vegetables before tossing into salad. You may want to use a food processor to shred vegetables such as bok choy and Chinese lettuce. Small cubes of regular or firm tofu or toasted slivered almonds will also boost the calcium.

SANDWICH STUFFING

Lightly steamed and cooled bok choy, Chinese lettuce or high-calcium greens can be shredded (use food processor if you have the appropriate attachment) then stuffed into pita pockets along with your favorite spread.

BURRITOS

One of the fastest ways of making a calcium-enhanced sandwich is by stuffing a soft, round tortilla shell with canned refried beans and shredded soy cheese. Spread 1/4 cup (50 mL) to 1/2 cup (125 mL) refried beans over half of the tortilla shell, sprinkle 3 tbsp (45 mL) shredded soy cheese over top, then add some finely chopped tomatoes. Roll until completely wrapped. Heat in oven 5 to 10 minutes until cheese melts. Serve hot.

TACOS

Make a calcium-loaded taco by adding cooked, mashed high-calcium beans (see values for beans in the Calcium Table, page 174) mixed with prepared taco sauce. Add lightly steamed, cooled and shredded kale, soy cheese and chopped tomato. Include some hot sauce for an added kick. Remember, taco shells made with lime water gives an added boost of calcium. Try warming the taco shells in a 185° F (85° C) oven for 5 minutes before stuffing them.

LASAGNA

Add cooked quinoa, high-calcium green vegetables and/or mashed beans between the layers of soy cheese, noodles and prepared tomato sauce.

PUDDING

To an infant, nothing is more satisfying than a sweet banana. For added calcium, try supplementing with soft, fresh figs: Cut 3 figs into tiny pieces; mash with a fork and add to 1 mashed ripened banana. A mini-blender is helpful for creating a smoother consistency.

CAKE ICING

In a blender combine 1/2 cup (125 mL) regular tofu and 1 cup (250 mL) store-bought icing. Blend until smooth and spread on cake or muffins.

ROASTED NUTS

Toasted nuts are delicious as a snack. They are also wonderful as a garnish or thrown into a salad. Try almonds, hazelnuts, brazil nuts and pistachios. Preheat oven to 350° F (180° C) and roast nuts up to 10 minutes. Nuts should be stirred in the pan every 3 minutes. When roasting chopped nuts, reduce roasting time 3 to 4 minutes. Toss with a little oil and season to taste.

CRUMBLY NUT TOPPING

Add a calcium-rich crumb topping to pasta, rice, a vegetable stir fry or other meals. Simply heat 1/2 tsp (2 mL) oil in a nonstick pan. Sauté 2 cloves minced garlic for a few seconds then add 1/4 cup (50 mL) bread crumbs, 1/2 cup (125 mL) ground almonds and seasonings of your choice (salt, pepper, garlic powder or paprika or a combination of spices); sauté, stirring often, for an additional 3 to 4 minutes. You can substitute almonds for hazelnuts or pistachios.

SUPER SALAD DRESSING

Boost the calcium of almost any store-bought salad dressing by simply adding 6 oz (175 g) regular or soft tofu to 1 1/2 cups (375 mL) salad dressing. Mix in blender at high speed until creamy.

APPENDIX B
Nutritional Notes on Calcium

Calcium intake recommendations

North American calcium recommendations are designed for conventional eaters. Typically, the recommended amounts are high — ranging from 800 to 1000 mg per day — and are based on the typical "western diet," with its heavy emphasis on protein and salt-laden foods. Such high amounts are not surprising, since excess protein and excess salt tend to draw calcium out of the body through the urine. People moving away from this style of eating, and towards a greater emphasis on fruits, vegetables, whole grains and bean products, are thought to require far less calcium than conventional recommendations suggest. It has been frequently noted that various population groups around the world are able to maintain healthy bone mass with much lower daily calcium intakes. In fact, the World Health Organization (WHO) recommends substantially less calcium (400 to 500 mg) than is typical in North America. Thailand's recommendation, for example, is 400 mg per day, while Japan and Korea recommend 600 mg per day.

Calcium absorption

The calcium contained in some foods is more easily absorbed by the body than it is in others. For example, calcium in green vegetables is thought to be better absorbed than that contained in dairy products, whereas calcium from dry beans (legumes) may not be absorbed as well as that contained in vegetables. It is difficult to make generalizations, however, since there is no calcium-absorption data available for many plant foods. There are also other factors that affect calcium absorption, such as vitamin D intake (including exposure to sun). Finally, plant-based foods can be high in calcium but also high in *oxalic acid* — examples include spinach, chard, and rhubarb, all of which have been excluded from our Calcium Table (see page 174) for this reason.

Calcium content varies

The New Laurel's Kitchen cookbook provides the following excellent advice about measuring the nutrient composition of foods: *"It is not that food scientists can't measure nutrient levels accurately; they can. But a great many other factors affect whether the apple in your hand will contain the nutrients listed in USDA Handbook 8 under 'Apple, raw, medium, 3-inch diameter.' Weather conditions while growing, soil type, soil minerals, fertilizer used (if any), variety, water source, storage conditions, and many other factors can significantly alter nutrient levels."* (Excerpted from *The New Laurel's Kitchen*, copyright © 1986 by The Blue Mountain Center of Meditation, Inc., with permission from Ten Speed Press, P.O. Box 7123, Berkeley, California 94707.)

Pregnant and nursing women

It's commonly recommended that pregnant women and nursing mothers need more calcium than others. For more information, contact your public health department, a qualified health professional, visit the websites of the Toronto Vegetarian Association at www.veg.on.ca or that of The Vegetarian Resource Group at www.vrg.org.

Chinese leafy green vegetables

There is considerable confusion about the differences between various Chinese leafy green vegetables. And there tends to be an incomplete body of nutritional information on these wonderful foods, which are used extensively in traditional Chinese cooking. *Chinese cabbage* is often confused with *Chinese lettuce* (or *Long Napa*). From the same family of vegetables, we get *bok choy*, or *pak choy*, which contains 157 mg of calcium per cooked cup (250 mL). Other calcium-rich Chinese leafy green vegetables to look for: Chinese broccoli (*gai lohn*), baby bok choy (*nau bok choy*) Shanghai bok choy (*green baby bok choy*) and yu choy. *Note: The information on leafy greens above has been reviewed by Judy Chong of Wing Chong Farm Ltd., Toronto, Canada — respected growers and distributors of Chinese vegetables. For information and photographs of fresh produce, browse the Wegmans Supermarkets website www.wegmans.com.*

Cooked versus raw

Cooked vegetables tend to lose volume. That's why our Calcium Table often indicates that cooked vegetables contain more calcium *per measure* than raw.

Fresh versus frozen

Frozen vegetables tend to be compacted during the packaging process, with the result that they often contain more product per measure than fresh. With more product per measure, you eat more and therefore will consume more nutrients — including calcium.

Calcium in tofu

Although our Calcium Table (page 174) lists the calcium content for tofu, in reality the calcium content of tofu varies considerably. Look for tofu processed either with "magnesium chloride" or "calcium sulphate." For an excellent article on the calcium content of tofu, request the *Nutrition Hotline* article by Suzanne Havala, M.S., RD, in the July/August 1997 issue of *Vegetarian Journal*, published by The Vegetarian Resource Group, at vrg@vrg.org, or link up to their website at www.vrg.org, or write them at P.O. Box 1463, Baltimore, Maryland 21203.

Fiber and calcium

It is sometimes said that a high-fiber diet is not good for calcium absorption. This flies in the face of not only every major health recommendation to eat a high-fiber diet (rich in plant-based foods), but also does not stand up to population studies that show fiber-rich diets do not appear to impact calcium absorption. On the contrary, a diet high in calcium-rich plant foods (such as those provided in this book) are an excellent way of obtaining calcium — while also getting the advantages that plant foods offer over animal products.

Fortified non-dairy "milk" beverages

Fortified non-dairy beverages contain amounts of calcium comparable to cow's milk. (See http://www.soyfoods.com/nutrition/CalciumChart.html., and our own Calcium Table [next page]; also see *Guide to Non-dairy "Milks"* in the Jan/Feb 1998 issue of Vegetarian Journal at www.vrg.org/journal). Fortified non-dairy beverages — such as soy milk — have long been available in the U.S. but not in Canada. The Canadian government department, Health Canada, has recently lifted this ban. (See following.)

Non-dairy calcium beverages in Canada

Before December 1997, calcium-fortified non-dairy beverages were not allowed into Canada. With the new regulations from Health Canada announced in November 1997, Canadians have begun to see fortified non-dairy beverages to which their American friends have long had access. Canadians can now find a variety of non-dairy fortified soy "milks" in the dairy case of their local supermarket.

Sesame products

Sesame products are often thought to be a good source of calcium. Although whole sesame seeds with the hulls (shells) are very high in calcium, the hulls are also thought to be high in oxalic acid, thus possibly interfering with calcium absorption. Much more common are hulled sesame seeds and tahini, both of which have the hulls removed and are typically the ones you find in grocery stores. Consider the thoughts of dietician Suzanne Havala: "...the calcium content of tahini varies, and the amount you absorb will depend upon multiple factors anyway. Look at tahini as being a reasonably good source of calcium..."

APPENDIX C
Calcium Table
Approximate Calcium Values from Various Plant Foods

VEGETABLES

Beet Greens	1 cup (250 mL), raw	45mg	1 cup (250 mL), cooked	164mg
Bok Choy	1 cup (250 mL), raw	73mg	1 cup (250 mL), cooked	157mg
Broccoli	1 cup (250 mL), raw	42mg	1 cup (250 mL), cooked	72mg
Brussels Sprouts	1 cup (250 mL), raw	37mg	1 cup (250 mL), cooked	56mg
Carrot	1 medium, raw	15-20mg		
	1 cup (250 mL), raw	34mg		
Chinese Cabbage (Napa) or Chinese Lettuce (Long Napa)	1 cup (250 mL), raw	59mg	1 cup (250 mL), cooked	158mg
Collard Greens	1 cup (250 mL), raw	52mg	1 cup (250 mL), boiled	152mg
Dandelion Greens	1 cup (250 mL), raw	103mg		
Kale	1 cup (250 mL), raw	93mg	1 cup (250 mL), boiled	163mg
Mustard Greens	1 cup (250 mL), raw	57mg	1 cup (250 mL), boiled	103mg
Okra	1 cup (250 mL), raw	81mg	1 cup (250 mL), cooked, fresh	100mg
			1 cup (250 mL), cooked, frozen	176mg
Parsnips	1 cup (250 mL), raw	48mg	1 cup (250 mL), cooked	57mg
Rapini (Broccoli Raab)	1 cup (250 mL), raw	137mg		
Rutabaga	1 cup (250 mL), raw	66mg	1 cup (250 mL), cooked	115mg
Snap Beans (Green) and Wax Beans (Yellow)	1 cup (250 mL), raw	40mg	1 cup (250 mL), cooked	57mg
Squash, Summer				
Zucchini	1 cup (250 mL), raw	18mg	1 cup (250 mL), cooked	23mg
Crookneck and Straightjacket	1 cup (250 mL), raw	27mg	1 cup (250 mL), cooked	48mg
Squash, Winter				
Acorn			1 cup (250 mL), cooked	64mg
Butternut			1 cup (250 mL), cooked	84mg
Hubbard			1 cup (250 mL), cooked	24mg
Sweet Potato (yam)	1 medium, raw	32mg		
Turnip Greens	1 cup (250 mL), raw	104mg		

BEANS/LEGUMES (COOKED)

Adzuki Beans	1 cup (250 mL)	64mg		
Black (Turtle) Beans	1 cup (250 mL)	102mg		
Garbanzo Beans (Chickpeas)	1 cup (250 mL)	80mg		
Great Northern Beans	1 cup (250 mL)	120mg		
Kidney Beans	1 cup (250 mL)	86mg		
Lima Beans	1 cup (250 mL)	53mg		
Navy Beans (Pea Beans)	1 cup (250 mL)	127mg		
Pinto Beans	1 cup (250 mL)	82mg		
Refried Beans	1 cup (250 mL), canned	116mg		
Soybeans	1 cup (250 mL)	175mg	1 cup (250 mL), roasted	235mg
White Beans	1 cup (250 mL), small	130mg	1 cup (250 mL), regular	160mg
Tofu, extra firm	1/2 cup (125 mL)	270mg		
Tofu, firm	1/2 cup (125 mL)	257mg		
Tofu, regular	1/2 cup (125 mL)	130mg		
Tofu, soft	1/2 cup (125 mL)	195mg		

Fruits

Currants, black	1 cup (250 mL)	61mg
Figs (dried)	1 medium	27mg
	1 cup (250 mL), about 12	286mg
Orange	1 medium	48 - 60mg
Papaya	1 medium	73mg
Raisins, seedless	1 cup (250 mL)	82mg
Kiwi	1 cup (250 mL)	20mg

Nuts and Seeds

Almonds	1 cup (250 mL), ground	252mg
	1 cup (250 mL), whole	377mg
	1 oz, whole (24 nuts)	75mg
Almond Butter	1 tbsp	43mg
Brazil Nuts	1 oz (25 g); 6 to 8 nuts	50mg
Hazelnuts (Filberts)	1oz (25 g), dry roasted,	55mg
Pistachio Nuts	1 oz (25 g), fresh (47 nuts)	38mg
	dry roasted	20mg
Sesame Seeds	1 oz (25 g)	
(whole seeds: roasted, toasted)		280mg
(Regular — kernels only, toasted)		37mg
Sesame Butter		
(Tahini)	1 tbsp (15 mL)	
	(from roasted kernels)	64mg
	(from unroasted kernels)	21mg

See Calcium Notes Appendix B

Sunflower Seeds	1 oz (25 g), hulled	19mg
Walnuts,		
English or Persian	1 oz (25 g)	26mg

Flour

Soy flour	1 cup (250 mL), full-fat	160mg
	1 cup (250 mL), defatted	240mg
Amaranth flour	1 cup (250 mL)	184mg

Whole Grains (cooked)

Amaranth	1 cup (250 mL)	298mg
Quinoa	1 cup (250 mL)	102mg

Dry Seaweed

Arame, Hijiki,	1 oz (25 g)	375-450mg
Wakame		

Miscellaneous

Carob powder	1 tbsp (15 mL)	28mg
Maple Syrup	1/4 cup (50 mL)	53mg
Molasses		
(Blackstrap)	1 tbsp (15 mL)	176mg

Fortified Non-Dairy "Milks"

1 cup (250 mL)

DariFree (potato-based)	275mg
Edensoy Extra	200mg
Health Valley Soy Moo	400mg
Harmony Farms Enriched Rice Drink	400mg
Pacific Foods Multi-Grain	150mg
Pacific Rice Fat Free/Lowfat	150mg
Pacific Ultra-Plus/Lite	300mg
Rice Dream Enriched	300mg
So Good, Soya	280-300mg
So Nice, ProSoya	230mg
Solait	300mg
Sovex Better Than Milk light	500mg
Sovex Better Than Milk	350mg
Soy Dream Enriched	300mg
Vitasoy Enriched	300mg
Westsoy Plus	300mg
Westbrae Natural Rice Drink	250mg
Westbrae Oat Plus	300mg
White Wave	300mg

Other Beverages

1 cup (250 mL)

Calais (carbonated flavored drink)	235mg
Carrot juice, fresh	60mg
Fortified orange juice:	
Minute Maid Calcium Rich	300mg
Tropicana Pure Premium & Calcium	350mg

Principal Sources

U.S. Department of Agriculture, Agricultural
Research Service 1996. USDA Nutrient Database
for Standard Reference, Release 11.
Nutrient Data Laboratory Home Page,
http://www.nal.usda.gov/fnic/foodcomp
1997 Canadian Nutrient File

Other Sources

Food Values, Jean A.T. Pennington, 15th edition
Info Access (1988) Inc.
The New Laurel's Kitchen, by L. Robertson,
C. Flinders, B. Ruppenthal

APPENDIX D
Lifestyle Factors &
Osteoporosis

We're commonly advised to consume a lot of high-calcium dairy products to help prevent osteoporosis. Conventional wisdom suggests that we consume up to 1500 mg or more of calcium per day. But does the evidence support this view? Consider the following:

Fact   **Populations in less developed countries with much lower calcium consumption suffer very low rates of osteoporosis.**
Evidence   Researchers went to South Africa to look at the women of the Bantu peoples, who consume an average of only 440 mg of calcium per day and who breast-feed an average of 10 children over the course of a lifetime. Conventional thinking might conclude that these women would have serious bone weakness. In actual fact, these women were shown to have a very low incidence of osteoporosis. A major study recently undertaken in Gambia has yielded similar findings. Many health professionals feel that an active lifestyle and a lower intake of animal protein and salt-laden foods offers the best protection against calcium loss.

Fact   **Populations where consumption of animal protein is highest (wealthy nations) also suffer from the most osteoporosis — even though they also tend to consume the most calcium.**
Evidence   People in western countries suffer from the highest incidence of osteoporosis. We also consume the most calcium, as well as the most animal protein. One non-western population group that suffers from osteoporosis is the Inuit (Eskimos). In the early 1970s, researchers discovered that even though the Inuit consumed a diet very high in calcium (2000 mg per day, principally from animal bones in their fish), and also ate 250 to 400 grams of animal protein daily, their incidence of osteoporosis was very high.

Popular opinion suggests that osteoporosis is a disease of calcium deficiency. The largely untold flip side to this story is that calcium loss plays a major role in the onset of osteoporosis. Many health professionals point out that by attempting to consume a lot of calcium, without regard to a number of other lifestyle factors, will not likely compensate for calcium loss. Therefore calcium balance — the relationship between calcium intake and calcium loss — determines how much calcium is left remaining to be used by our bones.

It is far better to try to prevent osteoporosis than to have to treat it later. And it's equally unwise to concentrate merely on calcium intake and forget about a number of important factors in the whole bone-strength equation. (For a complete discussion on osteoporosis and lifestyle, we'd recommend *Better Bones, Better Body*, Susan E. Brown, Ph.D [Keats Publishing, 1996].)

The following lifestyle choices can help prevent osteoporosis:

1) Lower your intake of salt and animal protein
Conventional sources of information often gloss over the fact that eating too much salt and protein (which we do in western societies) is regarded by many health professionals as being major factors in calcium loss.

Excess protein (particularly from meat, fish and poultry) and too much salt (particularly from processed foods) washes calcium out through the urine. In the case of animal protein, it's understood that animal products (such as poultry, beef, fish, etc.) contain high amounts of sulfur and phosphorus which, as with salt, has the effect of forcing calcium out of the body — from the kidneys and out through the urine. A doubling of protein intake is thought to increase calcium losses by about 50 percent. This same process applies to children as it does adults.

Trying to reduce the amount of salt in our diet can be confusing. Conventionally, 75% of the salt we eat is hidden in processed foods — everything from breakfast cereals, processed cheese, canned soups and frozen dinners. Only 15% of our salt intake comes from the salt we add with the shaker, while about 10% comes from foods in their natural state.

We have no nutritional need to add salt to foods. We only need a few hundred milligrams of salt in our diet daily. Many of us are eating between 4000 mg and 6000 mg. Some of us are consuming up to 10,000 mg daily.

When you must add salt to your diet, use low-sodium products such as low-sodium soy sauce or low-sodium soup mix.

According to Suzanne Havala, M.S., RD, "...it's very likely that if you moderate your [animal] protein intake and limit your intake of sodium and salty foods, your calcium needs will be lower than those of the typical [person]." How much lower? According to respected calcium researcher Dr. Robert P. Heaney: "On diets low in sodium and protein, it's possible to show that the calcium requirement could be as low as 500 milligrams per day." At a 1995 calcium symposium, he is reported to have said much the same thing: that on low-protein, low-salt diets, our calcium requirement "may in fact be as low as 400 to 450 mg/day..." At the same conference, this viewpoint was supported by fellow calcium researcher Dr. B.E. Christopher Nordin, who is quoted as saying: "If your urinary calcium is [very low] and your calcium absorption is high, you can manage on a few hundred mg of calcium [per day]."

If you've heard that animal protein is somehow critical to a healthy diet, many health professionals think otherwise. Vegetarians, for example, who tend to eat less (or no) animal protein, have been shown to retain more calcium than meat eaters (omnivores). There has been some research to suggest that animal protein forces calcium out of the body far more than plant protein does. It is even theorized by some health professionals that since dairy products are sources of animal protein, they may even contribute to calcium loss. Others, such as Cornell University professor of nutritional biochemistry T. Colin Campbell, Ph.D, argue that "animal protein is one of the most toxic nutrients of all."

2) Exercise

According to the Osteoporosis Society of Canada, "Regular and moderate exercise will help individuals to attain their highest peak bone mass, which provides a 'reserve' of strong bone, and ensures that bone loss does not become significant until much later in life." Brisk walking (preferably with hand weights) skiing, jogging, tennis or other forms of weight-bearing exercise (swimming is not weight-bearing) are all good choices. It's particularly important that young people be physically active, as bone mass reaches its peak around age 25. Find a form of exercise that works for you and stick with it. *Note:* If you have been inactive for some time, or if you have any medical condition that warrants caution, first consult with a qualified health professional.

3) Vitamin D

Vitamin D from sunlight helps the body to absorb calcium. Summer sunlight offers far more vitamin D than winter, while winter in southern climates is a better source of vitamin D than in northern locations. Those living in northern locations or having little opportunity for direct sunlight exposure, may wish to consider adding a vitamin D supplement to their diet, or consume fortified non-dairy beverages containing vitamin D. Cow's milk contains vitamin D only because it is added. People over 50 are thought to be particularly susceptible to vitamin D deficiency. For those over 50, the U.S. National Academy of Sciences recommends 400 IUs of vitamin D per day — 600 IUs for those over 70.

Sun-protection lotion has been shown to interfere with the UV rays from sunlight that stimulates vitamin D absorption. According to the Osteoporosis Society of Canada, sunscreen lotion with an SPF factor higher than 8 reduces our ability to produce vitamin D. It is wise, of course, not to get too much sun, as this is thought to lead to skin cancer.

4) Calcium intake

Conventional recommendations suggest we need high intakes of calcium — 1000 mg per day or more. Alternative viewpoints point out that people consuming less protein and salt — and thus losing less calcium — require less calcium, although it's not known what would be the optimal amount of calcium intake with a calcium-friendly diet and lifestyle. Calcium intake appears to be particularly important during the pre-adolescent and adolescent years, as young people appear to have an ability to absorb calcium more efficiently than adults, and because youngsters experience more rapid growth — including bone growth — than adults.

The "Recommended Dietary Allowances" (RDAs) for calcium and other nutrients were created to identify a safe and adequate amount of various nutrients for virtually all healthy people. The committees drawing up these

numbers have had to take into account that people come in all sizes, have significant variation in their needs based on their genetics, their age, and their other lifestyle habits — including exercise patterns, alcohol intake, salt intake, protein intake, and smoking patterns — to name a few. So the RDA committees have had a tough task — providing big "margins of safety" in their nutrition recommendations. They do this by setting their recommendations two standard deviations above the average requirement. This simply means that for 98% of the population the RDA for any given nutrient (except calories) is actually set higher than the requirement. (For a complete discussion of the principles and assumptions behind the development of the American RDAs, see: chapter 2, "Definitions and Applications" in the 10th edition of *Recommended Dietary Allowances*, published by the National Research Council, Food and Nutrition Board Subcommittee.)

5) Don't smoke
What some people do not yet appreciate about smoking is in just how many ways it is bad for your health. Smoking is known to impair metabolism, slow wound healing, and expose all body tissues to a large number of toxins. It is little wonder then that smoking accelerates the aging process, reduces fitness levels, and impairs normal metabolic processes that preserve bone strength. The more you smoke, the more likely you will have weakened bones. A 1997 study published in the *British Medical Journal* suggests that one out of every eight hip fractures can be attributed to smoking.

6) Limit alcohol, cola, coffee, and sugar
For various reasons, these four items are thought to have a negative effect on our body's calcium supply. Sugar and caffeine together, as in cola beverages and coffee, is thought to be particularly damaging to bone health.

7) Be cautious with medications
Cortical steroids, thyroid drugs, diuretics, antibiotics, aluminum-containing antacid tablets, and others, are thought to possibly cause some calcium loss. Check with your healthcare professional.

Index

C

Cake, zucchini carrot, 128

Candied almonds, 135

Cappuccino, 158

Carrots:

    bean soup, 50

    and bok choy stir-fry, 85

    broccoli salad, 55

    broccoli soup, 42

    Brussels sprouts soup, 41

    coleslaw, 58

    juice, 163

    kale apple juice, 164

    with parsnips Moroccan-style, 89

    and zucchini cake, 128

Casserole, greens and grains mushroom, 68

"Cheesecake", peanut butter marble, 130-31

Chickpeas:

    and bean burgers, 107

    chili, 70

    falafel balls, 113

    hummus bean blend, 27

    lemon soup, 47

    penne with tomato sauce, 120-21

    spiral pasta salad, 57

    stew, 82-83

    sweet potato patties, 110

Chili, 70

Chinese lettuce:

    corn chowder, 48

    layered mashed potatoes, 100

    lemon ginger stir-fry, 84

    perogies, 78-79

    shepherd's pie, 74-75

    with slivered almonds and caraway, 90

    sweet potato patties, 110

    yellow bean curry soup, 45

Chocolate:

    hot, 157

    ice cream, 143

    malted banana smoothie, 151

    milk, 156

    mousse, 139

Chocolate chips:

    cookie pie, 134

    sesame cookies, 126

Chowder, corn, 48

Cilantro:

    bean salad, 53

    dressing, 55

Cocoa pancakes, 146

Coconut:

    almond cookies, 127

    almond milk, 160

Coffee, cappuccino, 158

Coleslaw, 58

Collard greens:

    -apple juice, 162